SCIENCE WITHOUT LIMITS

SCIENCE WITHOUT LIMITS

Toward a Theory of Interaction Between Nature and Knowledge

JAMES S. PERLMAN, Ph.D.

Prometheus Books

59 John Glenn Drive
Amherst, NewYork 14228-2197

Published 1995 by Prometheus Books

99 98 97 96 95   5 4 3 2 1

Library of Congress Cataloging-in-Publication Data

Perlman, James S.
    Science without limits : toward a theory of interaction between nature and knowledge / James S. Perlman.
        p.   cm.
    Includes bibliographical references.
    ISBN 0-87975-962-3 (hc : alk. paper)
    1. Science—Methodology. 2. Science—Philosophy. 3. Objectivity. I. Title.
Q175.P38276   1995
502.8—dc20   95-19752
                                                                                            CIP

Printed in the United States of America on acid-free paper

Partial Pictures

The musical notes are not the sound,
The map is not the land, the people, the culture.
Symbols are not the actuality
Any more than the smiling portrait is the person.

But beyond symbols, partial pictures are the best we get
Of earth, sun, stars and all things.
And when with other humans we interact,
We still deal with partial pictures.

Mutually, parent and child, wife and husband,
Acquaintance or friend, whoever it may personally be,
Casual or intimate, it is still images with which we deal
Sometimes beautiful, sometimes distorted, seldom with background, never complete.

But does Jack Know Jack, let alone Jill,
Whether seemingly spontaneous, reflective or provoking?
Is it people or *images-we-form* to which we react,
However we may label friends, foes, even ourselves, caringly or with hostility?

Encapsulated we are.

James S. Perlman

Contents

7

Preface

What role does the "human factor" play in the scientific process? How does it influence discovery? This volume examines the role of the scientist in the process of understanding the world. It emphasizes the interactive human character of scientific inquiry, a dynamic interplay between human beings and their surroundings which embodies attempts to understand, anticipate, and cope with nature. Scientists, ancient or modern, are seen as a conscious part of nature, interacting both physically and mentally with the world, forming an open system with what is observed, measured, and explained. This book addresses the early search for natural order, recognition of rhythms in nature, the use of guiding faiths or assumptions, building conceptual models and projecting them upon nature, feedback from nature, perpetual modification and reorganization of ideas, and the eventual recognition of interlocking scientific worlds. It also addresses scientific survival techniques, learning from experience, natural causes in the scientific process, conceptual images, and science and society.

All of this necessitates a reexamination of scientific objectivity, conceptual model building, the place of scientists in the hierarchy of natural systems, and a recognition of the inevitability of perpetual revolutions. Emphasis here is upon the *interactive* human character of scientific inquiry. The scientist approaches his surroundings using everything he has: senses, mind, muscles, intuition, and imagination. He is not a fully detached observer as he perceives, forms conceptual models (or images), and projects them upon the world to anticipate events. All this necessitates, as we shall see, a reexamination of scientific objectivity, of scientific conceptual model building, and of the scientist's place in the hierarchy of natural systems as well as recognition of the inevitability of perpetual revolutions in science.

The approach here is historical. The main topics are the early search for natural order, the recognition of rhythms in nature, the use of guiding faiths or assumptions, building conceptual models, projecting such models or images

9

upon nature, feedback, perpetual modification, and the reorganization of ideas and eventual recognition of interlocking (but tentative) worlds within worlds.

The following are the main developmental stages in this historical process:

1. *Survival techniques and learning from experience.* Science, a cultural heritage, originated in human efforts to survive. Through tools and techniques, men extended their bodies, minds, and imaginations to come to grips with the stern realities of nature. Man survived: Learning from experience was an early scientific achievement dating back to prehistoric times.

2. *Natural causes.* Nature was explicable in terms of the regular workings of nature, as in ancient Greek concepts of matter, rather than through supernatural intervention.

3. *Conceptual models or images.* Also, in searching for order, our ancestors partly created the order for which they were searching. They devised conceptual models, whether as constellations, calendars, zodiacal belts, or earth-centered spheres. In projecting conceptual models or images upon the heavens, early scientists shaped their data much as sculptors shape clay. As we do today, they were attempting to call the world to order with interactive senses, minds, and imaginations. Projected models explain observations and provide a framework for measurement. They are products of interaction, not of complete detachment.

4. *Successful prediction of significant specific events based on associated natural regularities.* A familiar example is the early Egyptian association of the heliacal rising of a star with the yearly flooding of the Nile. Such recognition and use of associated regularities of nature, indirectly and then directly, were forerunners of ad hoc hypothesis testing. This recognition of yearly regularities led to the development of the yearly calendar.

5. *Struggle for survival of scientific conceptual models.* An early example would be the gradual displacement of the monthly calendar by the yearly calendar based on such criteria as agricultural and commercial advantages. We also find here an early illustration of reorganization of scientific models. Later displacement of earth-centered models by sun-centered models of the heavens further illustrate perpetual revolution in science.

6. *Equations as ad hoc hypotheses for prediction.* Once Johann Kepler, Galileo Galilei, and Isaac Newton used equation relationships as hypotheses to be tested by comparing observed results to predicted results, scientific inquiry reached new modern phases. This can be illustrated by the prediction of the as yet unseen planets Uranus, Neptune, and Pluto using Newton's equation $\frac{F=GMm}{d^2}$ (the law of universal gravitation).

7. *Science and society.* Science, like other human enterprises, does not exist in a vacuum. Anthropologists have long been aware of that. Study of the mutual impact between ancient science and society bears this out well. In today's complicated world, there is need for natural scientists, social scientists, and humanists to work for a science of man. Man is a worthwhile focus of life, science, society, and values.

8. *A move toward a science of man.* A science of man involves the following considerations:

a) The human personality or microcosm exists at various levels of activity that may be designated as subatomic, chemical, physical, biological, socio-cultural, and cosmological.

b) Internal activities at these various levels are interrelated and variations in activity at one level affect activities at other levels. These activities at various interrelated levels afford mechanisms adaptive for change in a world of change.

c) The internal activities, interreactions, and changes of the human microcosm can be understood and often predicted in terms of basic laws within and between the various levels of organization.

d) Each human organism and group externally interacts with natural and sociocultural surroundings that shape it and are shaped by it.

In any case, science is an enterprise of human beings, about humans as a part of nature, and we hope, for human beings. The extent to which science and technology become increasingly more for human beings is a social problem to be socially determined.

9. *An interaction theory of knowledge.* Scientific "knowledge is structured in the human mind" and imagination—knowledge with its data, principles, models, laws, and techniques. If so, is this structuring part of a larger process that includes an open-ended system of events, signals, perception, patterning, projection, prediction, perpetual reorganization of ideas, and broadening of perspectives? In this open-ended process, are not the senses, mind, imagination, intuition, muscles, and apparatus operating interactively upon events (facts) or rather upon light or other signals (data) from events? And are not sensations, ideas, images, models, natural laws and beliefs all at the heart of knowledge—product ingredients of interaction?

The above nine aspects or stages in the development of science are historically traced and analyzed as interactions with nature as can be seen from the chapter headings listed in the Table of Contents on pp. 7–8.

No person is an island. Waters of time and space lap up onto the shore. This book is buttressed by the effort of many creative people from far lands and times. From immediate perspectives, I owe much to my wife, Beatrice, who did all the typing and even learned to operate a computer to facilitate my efforts. I dedicate this book to her. Editors Steven Mitchell and Eugene O'Connor also did much to steer me through swirling editorial waters.

Part One
The Dawn of Science

1

Introduction: Science as a Human Enterprise

The universe as we know it is a joint product of the observer and the observed.

Teilhard de Chardin, *Phenomenon of Man,* 1950

Science is not a cut-and-dried proposition. It is not technology alone. Nor is it merely a set of equations by and for a small group of eggheads. It is a dynamic interplay between man and his surroundings, embodying his attempts to understand, anticipate, and cope with natural events. Science arose out of our efforts to survive, our natural curiosity, and our search for order in a seemingly capricious, hostile world. Like art or government, science is a human enterprise that reflects man.* Its strengths and limitations are those of its human originators. Knowledge and perspective of science in relation to other human endeavors are vital to the future of our culture.

Anthropologists are well aware that technology arose mainly from man's efforts to survive, whether he shaped flint and clubs for protection, built fires for warmth, or devised calendars to anticipate events. Tools and technology became extensions of man's body, mind, and imagination. The hammer, the wheel, and the engine are extensions of man's muscles. The telescope, radio, and microscope are extensions of the senses to perceive distant or minute objects. Calendars and computers are extensions of man's mind. And no science or technology would exist without human imagination.

Today's technology has changed the face of the earth. Skyscrapers and highways replace hamlets and wilderness; jet planes challenge the birds for the skies. And even now rockets reach for the moon, planets, and stars. But there are also H-bombs, population explosions, and radioactive fallout. in helping us cope with nature, is technology a gift-bearing Prometheus, or has it become

---

*The word "man" is used in the generic sense of *Homo sapiens.*

15

a Frankenstein's monster? Does the danger reside in technological advance or in a lack of social intelligence? Will man be dehumanized by automation and a highly structured society? What about population pressures and natural resources?

Ancient Babylonia, Egypt, Greece, and Rome were highly structured civilizations based on slavery. Medieval serfdom survived in Europe into this century. There were slaves in the United States a century ago. If slaves are no longer prevalent, it isn't that we are necessarily more moral but that our technology has allowed us to substitute machines for bodies. In that sense, at least, technology has had a humanizing influence.

Technology can help us increase labor productivity; relieve drudgery; prevent and heal sickness; control birth; extend life; and free ourselves for a life of dignity, enlightenment, and wisdom. Technology in the service of man can mean increased humanism. Men need not be slaves to machines; the problem of technological advance is primarily one of an accompanying social reorganization in accordance with human values.

But science is more than technology, more than an accumulation of facts, principles, and gadgets. It is also a dynamic investigation of nature in a challenging search for order. It is an open-minded, systematic, and critical approach to the universe that insists on evidence. Early questions put to nature led to facts and theories, which in turn led to further questions in a continuing, reciprocal process that ever extends our horizons. Modern investigation of atoms, chromosomes, and cancer cells; of the moon, planets, and stars; or of man himself reveals science as a human enterprise that constantly pushes the boundaries of the known into the unknown. In the dynamic search for order, an understanding of individual objects has developed that, unfortunately, does not yet extend to human relationships. A science of man is needed.

Men have minds and imagination as well as senses. They not only learn, adapt, and invent, they build systems of ideas as well as of things. Philosophy, mathematics, religion, literature, music, art, economics, politics, and social theory are some of the idea systems built by men. And, of course, we have houses, ships, and bridges when men build systems of things. None of these tangible products, however, is possible without the ideas behind them. Science is a synthesis of the world of ideas and of things, and man is the synthesizer. In astronomy, when men grouped stars into constellations and projected domes or spheres upon the heavens in order to locate moving objects there, they were scientifically system building. They were projecting ideas upon nature to systematize nature. More than that, they were interacting with nature: light signals from the stars moved inward; the celestial sphere was projected outward.

The heavens above are a vast natural laboratory filled with light, beauty, and mystery. And even the earliest men could see that celestial motions were generally periodic and regular. It wasn't that the sun merely moved, but that it did so daily across the same belt in the sky. It wasn't that the moon simply had phases, but that those phases exactly repeated themselves. Spring always followed winter; and summer, spring. Such regularities suggested order, natural law, and a system. They encouraged calendar and chart making; they evoked questions and gave birth to man-made conceptual models of the world. And, through the wonder and grandeur of it all, the same sky that stimulated men to develop systems of astronomy and mathematics gave birth to religious systems, poetry, and art.

In his interactions with nature, man, whether in everyday life, art, science, philosophy, or religion, uses his mind, imagination, and senses. But none of these faculties can be trusted alone. First of all, although indispensable, *our senses alone cannot be trusted.* Olives look larger in brine than on the tree; a spoon appears to bend when dipped in a cup of tea. The sun seems to move across the sky. A golf ball appears to have little, if any, empty space within it. Things are not always as they appear directly to our senses. Plato suggested, in his famous story of the cave, that our senses give us only flickering shadows of reality. According to Plato, we are like men in a cave at night with our backs to a fire at the cave entrance. All that we get of the outside world through our senses are flickering shadows on the cave wall before us. To Plato, the senses give us superficial, fleeting shadows of events rather than events themselves. He therefore minimized the importance of the senses and fell back upon the mind for knowledge of the world. Logic and ideas, not shadowy sensations, were to be depended upon for understanding nature.

But *the mind alone cannot be trusted.* Life is bigger than any system of logic, whether in philosophy, religion, science, or mathematics. So far, systems of logic have been like sieves through which natural processes somehow ooze and so escape. Also, every idea or instance of reasoning involves one or more assumptions, and therefore can be no more valid than its original assumptions. Every idea or system of reasoning is vulnerable at the level of assumptions no matter how internally consistent it may be. In arithmetic, for example, we learn that $3 + 4 = 7$, but actually, $3 + 4$ may $= 5$. If we walk 3 miles due west and 4 miles dues north, we are only 5 miles from where we started. The statement that 3 miles + 4 miles = 7 miles involves the assumption that direction or change of direction does not matter. The total walk is 7 miles regardless of direction. But if direction *is* important, then 3 miles + 4 miles geometrically gives only 5 miles, the actual final distance from the starting point. The statement $3 + 4 = 7$ does not cover all cases and therefore applies as logic only in special, arithmetical situations.

Similarly, *imagination alone cannot be trusted.* Unaided imagination cannot differentiate fact from fancy. Indeed, it can breed illusions and delusions, and is therefore often on the same side of the same coin as sensory illusions.

Since neither senses, mind, nor imagination can be trusted alone, man must use everything he has in a system of checks and balances. At least, that is the position of science in its basic method of inquiry, hypothesis testing. Forming a new hypothesis requires imagination (courage, too!). This was certainly true for the Copernican idea that the earth moves around the sun, particularly at a time when tradition as well as appearances favored an earth-centered universe. It also took imagination as well as observation and reasoning to propose that men evolved from simpler forms of life. But it requires additional reasoning and observation to test hypotheses or theories once they are formed. Generally, in hypothesis testing, events are predicted (that is, reasoned through in advance) from the hypothesis. Actual events are then observed and compared to the predictions. If the *observed* results *match* the *predicted* results, the underlying hypothesis is tentatively accepted; otherwise, it is rejected. Thus, hypothesis formation and testing generally involve imagination, reasoning, and observation in a system of checks and balances. It involves intuition, too! Science is thus a human enterprise in interaction with nature.

You have undoubtedly heard the story about several blind men who came upon an elephant. None of them had previous knowledge of or experience with an elephant, and each one, upon examining the animal, came in contact with a different part of its body. The first man touched the elephant's tail and decided that the elephant was ropelike. The second came in contact with its side and decided that the elephant was wall-like. To the third man, who touched the leg, the elephant was stumplike, whereas to the fourth, who felt the trunk, the object was snakelike. Partial pictures!

In our approach to reality, we are often like these blind men. We have limited ideas based upon finite knowledge and experience. Yet we are in the habit of projecting our limited concepts upon the universe and then identifying these as the final answers to the universe itself. We do not distinguish between our concepts of reality and reality itself. Our ideas become final and absolute instead of merely the best approximation of reality under the circumstances of limited senses, knowledge, tools, and experience. We constantly form concepts based upon that part of the elephant with which we happen to be in contact through our particular professional, religious, racial, or national groupings. If we are unwise, we call our own particular incomplete picture the final reality; we declare the tail an elephant and the elephant a rope. Encapsulation, indeed! The danger of premature absolutes is that we shut ourselves off from possibilities of additional knowledge, more complete pictures, and a closer approach to reality itself. Worse than that, we perpetuate professional, religious, racial, and national walls, and we set ourselves in motion against one another.

Science, too, has been caught in this failure to distinguish between concepts of reality and reality itself. We encounter instances in which even outstanding men of science too often show a disappointing lack of an open-minded, systematic, and critical approach to problems outside their own field. At the close of the last century, a group of prominent scientists meeting at an Eastern university resolved that the physical sciences had attained full scope in Newtonian mechanics. That is, Sir Isaac Newton's laws of motion and gravitation fully covered a universe in which an absolute space provided a fixed framework. A second was always a second; a mile always a mile. All that was claimed necessary was more and more precision, not revised structure or scope.

Consequently, in 1905, Einstein's special theory of relativity came like a bombshell, revealing that Newtonian mechanics is only part of a larger picture in which space, time, and simultaneity of events are not absolute and fixed, but relative to the observer; that metersticks shrink and clocks slow down when the motion of the system on which they are located increases; and that the mass or quantity of matter of any object increases as its velocity increases. The props of the formerly firm mechanics were shaken, holding only under limited conditions. The scientific community was embarrassed, caught short. Why? Their science, like other areas of human experience, had not sufficiently differentiated between concepts of reality and reality itself. It had projected its concepts of an absolute time and an absolute space upon the universe and had then identified its own concepts as realities or properties of the universe itself.

How then is science, or man, to bridge the gap between concepts and reality? The first step is to recognize that the difference exists, that on the one hand there is man with his senses, physical and mental tools, feelings, imagination, and the ideas based upon all of these; and on the other hand, existing independently of concepts, is the universe.

The second step toward a mature approach to reality is recognition that our senses, tools, experiences, knowledge, and consequently our concepts are of a limited, selective, relative character. First of all, we are limited to the particular receiving mechanisms of the sensory and mental apparatus that we possess. Reality and the universe are sending out their signals in many different ways, some known and some unknown to us. Waves would be one example. We catch only those messages or waves for which we have receiving mechanisms. For example, by virtue of our eyes and an accompanying nervous system, we receive waves in a range of about 3,500–7,000 angstroms to give us light, sight, and color. The range of frequencies involved here is a very narrow one in a very broad band of electromagnetic waves including, for example, ultraviolet, X-rays, gamma ray, infrared, and even radio waves. Yet our eyes are not delicate enough instruments to catch any wavelength longer than red light or shorter than violet, just as we are not aware of sound waves below or above a certain pitch or frequency. Yet these waves exist. Japanese experts in earthquake phenomena some time ago reported a sensitivity of animals to earthquake waves that human beings would ordinarily be unaware of except through instruments. Radio waves are constantly passing all around us; we are aware of them only when we set up the proper receiving mechanisms. Man has extended his senses through such tools as the radio and seismograph, but the number and diversity of these receiving mechanisms are as yet small. There should be many, many more to come. Most that we now have can become much more precise. Present-day tools do not give us information that later tools will. They involve and select only the particular types of messages and knowledge for which they are designed. Meanwhile, we form concepts based upon limited and partial knowledge, concepts that only begin to approach reality.

Further, if the nature of our knowledge depends on the types of senses and tools we have, then the nature of our tools depends on accumulated past knowledge in a selective process. A great deal of data on atoms and atomic power has been obtained through atom-smashing devices such as cyclotrons, bevatrons, or synchrotrons. However, these powerful tools of artificial radioactivity have been possible only because the previous discovery and knowledge of radium and other naturally radioactive substances pointed to their development.

Granted our recognition of the incomplete and selective character of our factual knowledge and, therefore, of our ideas or concepts of reality, we are still left with the problem of reconciling our ideas of reality with reality itself. How can we keep our ideas in line with reality as we attempt to approach it? How can we minimize being embarrassed by reality as we try to form ideas about it? "Operationalism" and "frames of reference" are terms that have arisen in a possible answer to this question. To these, I add here a third term, *interactionism*, i.e., the mutual interaction between man and nature in the scientific quest for knowledge. It is a concept that I believe eventually leads to an interaction theory of knowledge, to be articulated later.

Operationalism, as first emphasized by P. W. Bridgman, contends that concepts or ideas—for example, space, time, democracy—should be defined and qualified functionally, that is, in terms of the evidence behind the concept. Since evidence generally arises through observations, tools, processes, and practices, *these* must enter into the definition of an idea. That is, the definition of an idea should be a working definition. For example, one calorie is operationally defined as that amount of heat that will raise the temperature of one gram of water from 15°C to 16°C. Or, a year, as time, would be understandable in terms of one revolution of the earth around the sun. Concepts thus become defined in terms of actual processes that occur rather than absolute, fixed properties imposed upon things. Light becomes understandable in terms of waves when associated with diffraction gratings but, on the other hand, as corpuscles or photons when photoelectric cells are used. Perhaps waves and photons will be seen as different aspects of the same thing or else as parts of a larger picture when more knowledge becomes available. Meanwhile, as Bridgman stated it in his *Laws of Modern Physics*, "If experience is defined in terms of commonly shared and commonly verified experience, there will always be correspondence between experience and our description of it." And meanwhile, prematurely set *a priori* principles need not exist to limit possibilities of the new experiences and new knowledge that may give the larger picture.

Then, for a mature, scientific approach to reality we might mention, thanks to Einstein, frames of reference broadly as that aspect of the relativity of knowledge arising from the particular background or system of motion in which an individual observer is located. If the year of time is determined by a complete revolution of the planet on which the observer happens to be, then a year to an observer on Mars would not be the same as to an observer on Earth; Mars is hardly halfway around its orbit when the earth is all the way around its own. Or again, if "up" is the direction away from the center of the earth, then, because the earth is spherical, to an observer in the United States "up" is in a direction in space nearly opposite to that for an observer in China. It depends on where the observer happens to be. And a bigger frame of reference may help: We are all on one spaceship, the earth, moving in space—who knows where?

We can say more about frames of reference in an extended anthropological sense. C. P. Snow emphasized two separate traditions or cultures, the literary and the scientific, and decried the chasm between them. To Snow's literary and scientific traditions, I would add a third, a *social* tradition, and would emphasize that in actuality, these three cultural traditions are parts of the cloth spun by man that in turn drapes and shapes him. These cultural drapings are frameworks of interaction. That is, man creates his traditions, which in turn influence him. The three cultural traditions, the literary, the scientific, and the social, have common origins in the makeup and interactions of men with both their surroundings and one another. For example, science insists upon *common evidence* for ideas and appeals to the senses that men have in common, whereas art bases itself upon *individual differences* in the apparatus, impressions, and insights of men. But both science and art are products of men, and both the common evidences of science and the individual differences of art are based on the nervous systems, minds, and senses that characterize human beings. Man has evolved from nature; and science, literature, and government from man. If man is a conscious part of nature interacting with the rest of nature in everyday life, science, art, music, or social theory, then none of these cultural products of man exists in isolation. Because they are interrelated, they affect and reflect one another in their developments and in man's development, too. A science of man is needed in the fullest sense of the words.

Galileo, in his advocacy of the Copernican over the Ptolemaic system of the universe and his difficulties under the Inquisition, brought into dramatic focus crosscurrents of the scientific, the literary, and the socioreligious. Since the time of Galileo, knowledge has multiplied and required increased specialization. This process has tended to narrow and deepen professional interests and to create artificial boundaries. More professional generalists are therefore necessary now within and between disciplines. This is already recognized in medicine. The internist serves the role of the professional generalist in medical practice. Life and personality are bigger than the immediate interests of the best specialists in the sciences, arts, or social sciences. Yet professionally, as otherwise, we are often like the blind men and the elephant, with limited ideas based upon finite knowledge and experience. We form concepts based upon that part of the elephant with which we happen to be in contact through our particular professional grouping. If we are not careful, we turn our own particular partial picture into a complete one. Work long enough upon the elephant's ear, and it may become the elephant to you. Specialists have much to gain through interdisciplinary sharing of experiences.

There is also a need for people in the natural sciences, social sciences, and humanities to work toward one another. A science of man is possible through common effort. And somewhere in the educational process, too, provision must be made to present things in their larger frameworks. Without specialization we are superficial; without perspective, provincial. Both specialization and perspective, analysis and synthesis, are necessary for an educated person.

Lastly, in a mature approach to our surroundings, let us further consider the interactive character of science and knowledge. If man is a conscious part of nature interacting with the rest of nature in science and otherwise, such interaction may afford insights not otherwise possible. What follows is meant to suggest an approach to a study of man based on interaction concepts and leading from the natural sciences. In this we take a cue from the idea of the ancient Greeks that each of us is a microcosm or miniature universe. But I emphasize that as microcosms we exist at various interrelated levels of activity, and that we are not island universes; we interact with other microcosms. We have surroundings that shape us and that are shaped by us. And as with the rest of nature, there is no artificial separation within us of physics, chemistry, biology, psychology, or sociology. We incorporate almost all the sciences, including the social sciences and the humanities.

To be specific, let us start with the following three propositions:

1. The human personality includes energy mechanisms adapted for change in a world of change on levels that may be designated as subatomic, chemical, physical, physiological, biological, psychological, social, and cultural.

2. The activities at these various levels are interrelated. Variations in activity at one level affect activities at other levels.

3. The internal activities, interactions, and changes of the human microcosm can be understood and often predicted in terms of basic laws within and between the various levels of organization, including the sociocultural.

These three propositions may be reinforced by such generalizations as the following:

1. The matter making up our bodies has the same electrical basis and is subject to the same subatomic, electrical laws as other objects around us.

2. The body in its locomotive, temperature-regulation, and sensory apparatus is equipped for dynamic physical relationships with a changing environment and is subject to universal laws of mechanics, thermodynamics, light, and sound.

3. The body through its own composition, cellular activity, and digestive, respiratory, and glandular systems performs chemical transformations for its own needs.

4. The cells, tissues, organs, nervous system, and brain with their specialized functions form a complex integration of activity for the well-being of the human organism.

5. Associated with the various activities above, biological, neurological, psychological, social, and cultural mechanisms condition the goal-seeking interactions of the human personality and society with its environment and, in turn, are affected by those interactions.

In sum, emphasis here is on nature, man-in-nature, and science as a product of man-in-nature. Main themes are man's scientific search for order, his guiding principles, his interaction with surroundings, his conceptual (and other) system building, his projection and testing of ideas. Perpetual revolutions in science result. In all this, the scientist is an interactor rather than a really detached observer. If so, scientific objectivity needs reexamination within a larger context of interaction rather than in the confinement of a supposed full detachment. All this has implications for reworking scientific definitions, concepts, theories, and laws as we shall see in the chapters to follow.

2

Ancient Science:
Survival Techniques and Learning by Experience

Science arose in the main from early human contact with nature and the necessity of anticipating certain of her events as a condition of survival.
Lloyd W. Taylor, *Physics, the Pioneering Science,* 1941

Science arose long before Copernicus, Galileo, or even ancient Egyptian surgeons. It arose in prehistory from survival interaction between human beings and their surroundings, when they began learning by experience in everyday survival and developing techniques to cope with the harsh realities of nature. It advanced whenever recognition of regularities in nature enabled anticipation of events. And it developed further when such learning and techniques were highly organized in such structured societies as ancient Egypt, Babylonia, China, or India. To be scientific, then as now, meant gathering ever more information about nature, changing ideas about nature, and improving techniques and tools in coping with nature. Our ancient ancestors, for example, expanded their knowledge and techniques as they moved from Old Stone Age nomadic food-gathering to settled village farming in a New Stone Age agricultural revolution. Whether gathering wild berries or later planting millet, they used everything they had—senses, mind, muscles, tools, accumulated knowledge and experience, curiosity, insights, and imagination—in interacting with nature and surviving. They planted and cultivated seeds in new places and nature produced food in these chosen places. The basic process was (and still is) one of human interaction with nature. Meanwhile, observation of plants and domesticated animals increased as did recognition of seasonal and other natural patterns. Time relationships between events became more apparent and some anticipation of events possible. The process was expansive.

Resulting population increases enabled New Stone Age cultures to build great urban centers at the mouths, and in the valleys and plains, of such mighty rivers as the Nile, Tigris, Euphrates, Hindus, and Yangtze. Huge public works involving large populations tamed the fertile valleys and plains for agricultural purposes. Large commercial centers also resulted as basic tools and weapons were increasingly fashioned from metals newly discovered, mined, and traded. Perhaps the constant increase of population was itself basic evidence that science existed in ancient times with expanding knowledge, techniques, ideas, and anticipation of events. Terrain, plant and animal life, and human society were entirely reshaped at the sites where cities grew up. All this should testify to the interactive human character of science at a basic level of survival techniques, learning by experience, social organization, and changed surroundings.

All this despite fanciful, mythological explanation! Yet at the time daily events and observed rhythms in nature at the time lay in the laps of the gods. Basic causes explaining events still were to be found in the realm and at the whims of-the supernatural. Witness the ingenious astrological ordering of sharp astronomical observations by the Babylonians as in their zodiacal belt and predictions.

In sum, science emerged in prehistory as successful human interaction with nature. That interaction involved a process and products characteristic of science. The products included an expanding body of knowledge, tools, and techniques. A distinguishing characteristic of the process is prediction or anticipation of events. More fully, the extent to which our ancient ancestors, in order to survive (and out of plain curiosity) (1) carefully observed events; (2) recognized relationships or patterns of events; (3) anticipated further events; (4) changed or expanded their techniques as they learned from experience; (5) transmitted information, skills, and experience by language, records, and current application; (6) socially organized for all the above; and (7) reshaped their surroundings for better survival—to that extent, they were scientific and their science interactive.

But now let us examine more closely known details of the Old Stone Age and later agricultural (New Stone Age) and urban revolutions of ancient Egypt and Babylonia to see to what extent these details satisfy our criteria for science listed above.

PREHISTORIC SCIENCE

The Old Stone Age people lived a nomadic existence, seeking food and other necessities where they could find them and then moving when local stores were depleted. Surroundings were primitive, wild. Shelters were improvised, temporary affairs, and the problem of sanitation was taken care of simply by moving on. Groups were small. Arts and crafts, very elementary, are limited by the use of simple stone tools. Flint, for example, was selected for cutting and scraping. Other rocks were selected and accepted for various uses as in making clubs, hammers, or diggers. Archaeological findings attest to all this.

Food gathering was limited to supplies available by hunting, trapping, fishing, root digging, or collecting wild fruit, grains, and grasses. Scraped hides and matted leaves and grasses afforded clothing materials. Boughs and mud enabled shelters to be constructed where there were no available caves.

Old Stone Age people did, however, possess skillful techniques in making and using elemental tools. They adapted, for warfare, hunting, and fishing, chipped flints, stones or bones for knives, scrapers, choppers, or weapons. They knew the methods of fire lighting and control for warmth and food preparation. They supposedly observed the regularity of the full moon for better night-time hunting, and the sun's changing noonday angles for seasonal digging and berry picking. There were stock breeding and migratory hoe gardening. Tool and weapon-making, in general, from the earliest of societies, meant determining by experience the best stones for different purposes, where they were to be found and how to be handled. Improved techniques developed with accumulated experience and knowledge, which was in turn handed down to new generations.

Besides transmission of information and techniques, there were many forms of social organization basic to successful scientific endeavors. Archaeological evidence indicates successful group hunts in long pursuits of migratory herds of deer, musk oxen, wild horses, and even mammoths and woolly rhinoceroses before and during the last Ice Age. For Stone Age inhabitants living tens of thousands of years ago, success necessitated common detailed knowledge of particular animals, their movements and habits, and the terrain. Success also involved team skills and cooperation among the hunters. The Old Stone Age still exists in nomadic areas inhabited by Bedouins, Yemenites, Pygmies, and other African tribes.

Cave drawings of animals reveal good physiological detail and observation. So do uncovered burial grounds and artifacts depicting living creatures and inanimate objects. Cave and burial sites also reveal a belief in magic. The arrow painted through the heart of the deer on the cave wall supposedly helped in the realization of a successful kill in the coming hunt. Causality unquestionably was at the mystical level, but that does not take the place of the careful observation, dexterity, skill, and experience—that is, the scientific ingredients—needed for a successful hunt. Nor does it deny these peoples' accumulated body of elementary astronomical, botanical, geological, and zoological knowledge.

We may further detail Old Stone Age life based on what we know or can safely infer. Transportation on water was by canoe, raft, or even boat; by beast of burden (e.g., camels, horses) on land; by sleds in snow country; and even by early wheeled contrivances in some areas. Communication was by spoken language and by demonstrated activity. Counting and enumeration also existed as the beginning of mathematics.

Prehistoric medicine, of course, had crude beginnings in "medicine man" herbs as well as in psychological effects induced by incantations and other mystical religious rites. Religion, art, and magic were meant, in part, to influence natural events symbolically through animal and human sacrifices, cave art, figurines of fertility, and other animistic or anthropormorphic means. (Lack of knowledge and of concept development in given areas had their encapsulations then as now. That is, it is not only what we know but what we *don't* know that narrows our thinking and partial pictures.)

But, to return to our main argument, on what basis may the development of arts and crafts as well as learning by pragmatic experience of Old Stone Age people be considered to lie at the heart of science? In order to answer that question, let us review Old Stone Age activities in terms specifically of the criteria for science enumerated already.

1. Expanded Information and Survival Techniques by Direct Observation and Experience

As already emphasized, Old Stone Age people acquired and tested information and techniques by direct observation and experience in their daily interaction with natural surroundings. If their immediate observations, reactions, and techniques had not been sharp, realistic, and generally effectual in dealing with wild animals, plants, rocks, and terrain in jungle, desert, snow fields, or on water, (physically) puny mankind could not have survived. But *Homo sapiens* did survive.

Food gathering, whether in the form of hunting, fishing, or digging, was facilitated by acquired knowledge of seasons and of their cycles and differences. This enabled early man to predict and plan, thus marking the beginning ot the scientific knowledge of climate and weather The lunar calendar based on the moon's phase cycles originated in the Old Stone Age and is still used to set religious holidays for Jews, Christians, and Moslems. As we shall see later, observational astronomy was perhaps the earliest and most sophisticated of ancient sciences.

Old Stone Age men invented the first actual "engine," the bow, through which human energy, accumulated in bow tensions, is released all at once. Poisoned arrows and spear tips involved basic botanical observations and technical skills. Harpoons, spears, bows and arrows, and other fishing and hunting devices illustrate elementary mechanical insights. Discernment between edible and inedible plants attested to basic botanical observation and techniques, as did herbal and other medical preparations. All these methods involved acute observation, thought, curiosity, and techniques directed to human needs and survival.

Fire, while a mystery and a terror when out of control, could be directed and controlled for human needs by applying rudimentary physical, geological, and chemical techniques. Fires were started by use of friction. Sparks were obtained, for example, by rubbing flint against iron pyrites or hematites. The correct rocks had to be found. In other cases, sparking was obtained by rubbing together two pieces of wood. Observation and skill were involved in fanning flames and feeding controlled fires.

Acquired knowledge and techniques were needed to make clothes. Stone knives, scrapers, and choppers were applied to furs, hides, and skins which had to be suitably shaped, pieced, and held together. It took skill and experience to fashion the stone tools effectively. More and more potters' wheels appeared for shaping clay pottery. With observation, thought, and resourcefulness, temporary shelters were improved by use of changed materials or new techniques. To further illustrate ancient resourcefulness and new techniques, some nomads eventually introduced hoe gardening as a beginning of food production to supplement the usual food gathering. We have archaeological evidence of the uses of symbols and reckoning in a prehistoric mathematics. Training horses, donkeys, camels, Eskimo dogs, and other animals for transportation required extended observation and knowledge of the animals as well as skill and techniques of training. Eventually goats and cows were husbanded for milk in some areas.

2. Cultural Transmission of Knowledge and Techniques

Old Stone Age people not only acquired but orally transmitted survival information, skills, and experiences to other group members. Common language definitely had a role in survival and in the development of techniques. And also very essential, of course, was cultural transmission of acquired knowledge and skills to young people by direct illustrative teaching, by apprenticeship, by relevant games, and by selective group practice in everyday life. Trade involved cultural exchanges of knowledge and techniques as well as of products with other groups.

3. Improvement and Testing of Tools and Techniques

Old Stone Age people also improved techniques previously acquired and transmitted. They tested them by application in their daily lives. Some illustrations of improved tools and techniques have already been given. Flint and other rocks, at first, were selected and used as is, for sharp cutting edges. In time, however, such rocks were further sharpened before using. Involved here again were observation, skills, thought, and accumulative experience that may be considered scientific for the following reasons: First, for a cutting purpose there had been operational discrimination among materials made of different stones as well as among materials of stone, bone, wood, ivory, shells, and eventually metals. Then came changing shapes or edges of materials in an adaptive modification of stone, bone, or other knives for greater cutting effectiveness. Thirdly, there was an expansion and testing in the *adaptive technique* itself. For example, instead of first selecting a sharply edged flint or obsidian stone to be reshaped for handling, a new process of flaking was developed. In this process, all pieces of stone material were chipped from a basic core piece to fully shape the knife and its edge as desired. Such flaking in turn required special tools. The new specialized tools were, of course, tested by successes of their end results; for example, sharper cutting knives were produced by core-flaking tools and techniques.

4. Recognition of Natural Cycles and Anticipation of Events

Old Stone Age people recognized cycles and other relationships in nature, thus enabling them to anticipate events and successfully organize future activities. Hunters, knowing the moon's cycle of phases, could organize night expeditions in anticipation of light from a full moon. The present-day Jewish calendar had its origin among ancient Hebrews who as desert nomads developed a calendar based upon the moon's phases. When the Hebrews arrived in the land of Canaan or Babylon (today's Iraq) and became an agricultural society, they found it necessary, along with the Babylonians, to reconcile their lunar calendar with the sun's yearly cycle through intercalations, i.e., inserting extra days. Precision in the observed lunar cycles and the calendars, however, remained high. While hunting based on the lunar cycle was no longer so vital, accuracy in naming days of special religious observances continued to be as serious as ever. The lunar calendar remained basic until the solar calendar was eventually wholly established by the Egyptians in their urban revolution, as we shall see.

In sum, early tools and techniques came clearly into view as extensions and applications of the human body, mind, and imagination in man's interplay with his surroundings for survival. Science thus had its dawn in prehistory. Scientific interaction with nature was at the basic level of development of survival techniques and of learning from direct experience in an everyday struggle for existence. The science lay primarily in careful observation, learning from experience, and anticipating events; the technology lay in the arts and crafts learned by experience. Interaction between man and nature existed not only in the observation, techniques, and learning already emphasized, but also particularly in the *feedback* (literally and figuratively) of nature as to what would and would not work. All this is in keeping with the main focus of this book on the development of science as human interaction with nature.

THE AGRICULTURAL REVOLUTION (7000 B.C.E.–)

The earliest approximate dates for the agricultural revolution are ca.7000 B.C.E. for the Far East and ca. 5000 B.C.E. for predynastic Egypt and Babylonia among other early cultures. An agricultural revolution occurs wherever and whenever—even today—a people go from a primarily nomadic food gathering to village-centered food producing. That is, from foraging, hunting, and fishing to farming and animal domestication.

Permanent farming settlements vitally changed living conditions. In regular crop cultivation, men plowed with oxen and water buffalo, replacing women with hoes in temporary garden plots. Seeds were now stored for future crops such as wheat, barley, lentils, check peas, onions, and flax in Egypt. Livestock was domesticated for food and transportation as well as for beasts of burden. Such animals included sheep, cattle, goats, swine, oxen, water buffalos, horses, camels, llamas, dogs, cats, and elephants. Hunting, fishing, and trapping became supplementary. Dwelling places and storage facilities were now more permanently constructed, which facilitated pottery making and cloth weaving. There was increased specialization of labor in expanding arts and crafts or in cotton cultivation. Larger inner group cooperation and exchange of products and knowledge resulted in variations in individual food production. Food gathering had been more of a common group effort. More self-sufficiency with extended cooperation in larger settlement groups meant improved survival and increased population. Productivity was increased; food supply was better controlled, and expanded trade was now possible with outside pastoral nomads. In turn, there was more need for larger families to assist in economic activities of common group benefit such as clearing forest land, draining marshes, and building settlement defenses and enclosures.

From a scientific viewpoint, permanent living conditions of an agricultural revolution extended and deepened observations of plant and animal life: There was closer contact with plant cultivation and with domesticated animals. With plant cultivation came detailed knowledge and observation of effective fertilizing, sowing, cultivating, harvesting, storing, and protection against disease. With animal breeding came knowledge of reproduction, food and water needs, milking, animal strengths and weaknesses, habits, diseases, harnessing, wool shearing, and the like. New skills, techniques, tools, and products accompanied new scientific knowledge in agriculture. Once again early man learned by and from experience along with expansion of scientific techniques and insights, and increased successful anticipation of natural events as with the birth of a calf. Fertilization and crop rotation were successful techniques that came from experience with soil exhaustion when nomads became settled farmers.

All such increases in knowledge of natural processes and skills meant increased control of food supply and improved chances of survival. They provided such additional food, clothing, and other products as cultivated grains, milk, more meat and poultry, wool, cotton, and flax in addition to skins and hides, mud bricks instead of boughs, and pottery for more durable vessels.

Plows adapted to beasts of burden largely replaced the garden plow for soil cultivation. Stone axes sharpened by improved grinding-and-polishing tools replaced axes of flaked stone or plain flint. Carts replaced sledges and animals replaced men in using wheels on wagons. With permanent settlements, the potter's wheel shaped clay into various forms that remained firm by firing, that is, by heating to over 1,000°F.

Elementary chemical interest existed in observed changes of color, texture, plasticity, and hardness in pottery. All pots from a given village would be uniform. Bricks were hardened from mud along the Nile, and Tigris and Euphrates rivers. Mixtures of clay, sand, water, and straw were molded and sun-dried. The loom created materials for garments in addition to the use of skins and hides. For improved storage facilities, silos and storage pits were lined with straw or matting. Sailboats appeared along the Nile.

Also to be emphasized from a scientific viewpoint was the further development of conceptual aids and models. Agriculture meant increased emphasis on sunlight, seasons, the time of year, and eventually the Egyptian yearly calendar for planning activities. Meanwhile, lunar calendars for anticipation of natural events, for the planning of farming and other human activities, and for religious celebrations were reconciled with seasonal cycles basic to agriculture.

A typical picture of Egyptian farming in its agricultural and urban periods is given in a May 1972 *National Geographic* article, concerning tomb paintings on "how grain was grown": "After seed was scattered over the Nile mud, sheep or cattle or donkeys, driven across the field, trampled the grain into the ground. When the crop ripened, inspectors came with measuring lines to set the government's quota. Farmers with sickles lopped off grain heads and oxen or donkeys threshed the grain with men adding the grain heads. Winnowers tossed grain with wooden scoops to separate kernel from chaff."[1] By this time, there apparently was larger-scale production.

SCIENCE AND TECHNOLOGY IN THE URBAN REVOLUTION (CA. 3000 B.C.E.–)

The urban revolution was an expansion from village-centered farming to urban-centered specialized production in highly structured societies. Early urban revolutions had unique geographic centers at the mouths of long rivers that flooded fertile valleys and deltas, whether the Egyptian Nile Delta or the

Tigris-Euphrates convergence and plains of Babylonia (Iraq). India, of course, had its Hindus River and China its Yellow and Yantze rivers. For our purposes, we shall confine ourselves to details of the urban revolutions in Egypt and Babylonia. It was not only the periodic flooding, silt, and natural irrigation at the mouths of the mighty rivers that gave such locations definite agricultural advantages. Proximity to the Mediterranean Sea and the Persian Gulf offered unquestionable trade advantages.

The huge tasks involved in harnessing powerful rivers for agricultural survival necessitated large public works. To drain marshes; reclaim swamps; clear thickets; eliminate wild beasts; build and maintain canal networks, drainage channels, and dykes—all required organization, larger concentrated populations, and specialized techniques. In turn, the extensive, more permanent fertility of soil resulting from such large-scale projects could and did support ever larger populations. A further stabilizing factor in survival was the enlargement of diet. For example, dates, figs, olives, and other wild fruits could be easily cultivated, preserved, and transported. Large-scale political controls accompanied the widespread need of individuals for community irrigation and other projects.

Agriculture expansion and social reorganization led to technological needs for heavier-duty tools, conveyances, and weapons. Metals were discovered, mining and metallurgy developed. In the basic rise of sophistication in materials there was movement from stone to copper to bronze and eventually to iron (ca. 1400 B.C.E.). All this meant a developing knowledge of physical and chemical properties of metals in interaction with surroundings.

To discover and mine ores was a form of engineering. Salting copper, tin, or other ores certainly involved a functional chemistry, as did the production of bronze from copper and tin. Specialized metalsmiths and metal artisans shaped tools, weapons, and other products, including gold and silver art objects. More and more economic specialization was necessary, as was increased trade for metals and other needs. Egyptian *feluccas* with their triangular sails were ever more visible on the Nile and the Mediterranean. Improved boat construction in the use of these sails enabled transportation of heavier loads.

Fertile soil and labor specialization provided food surpluses that freed a large section of the population from food production for other specializations and for a highly structured autocratic society. The need for large-scale economic planning and administrative control gave an emergent aristocracy an absolute economic, political, and military power. Centralized planning and political-economic control were in the hands of a king and priests reinforced by nobility and warriors. The pharaohs of Egypt, aside from being chief priests, even made themselves into divinities. In Babylonia, the priesthood enjoyed even more power as direct agent-administrators of deities who supposedly owned all the land, basic wealth, and resources. As religious leaders and chief economic administrators with tax-collecting responsibilities, the priests in both Babylonia and Egypt were scribes and record keepers. They were astrologist-astronomers observing and interpreting divine messages presented in the relative motions of sun, moon, planets, and stars. They also became the healers, the physicians of the time. Established nobility gave military support and political advice to kings, political administrators, and the priesthood.

Beneath the top echelons in a descending hierarchy were warriors and surgeons; sculptors; farmers and fisherman; shepherds and camel breeders; and maintenance workers for rivers, canals, and other public enterprises. There were also merchants and, of course, sailors.

Among craft workers were miners and brick makers, smelters, metal workers and smiths, boat builders, spinners and weavers. Egyptians cultivated flax and wove linen as well. Winnowers tossed grain with wooden scoops to separate kernels of wheat from chaff. There were gem cutters, stone cutters, stone-hauling crews, glass makers, carpenters, gardeners, potters. Lowest of all were slaves generally for the dirtiest, heaviest, most dangerous tasks. (With the urban revolution, taking slaves in wars often replaced genocide.)

In short, urban revolutions promoted technical, scientific, and other specialization in survival interaction with surrounding. At the same time, specialization meant a structured interdependence of individuals and groups as well as of arts and crafts under autocracies or theocracies. Let us now consider the urban revolutions in Egypt and Babylonia in further detail for more specific scientific features. By "scientific" for those times, we again refer to direct learning from everyday life experience, testing ideas and techniques, whether the learning was initiated by curiosity, survival necessities, trial-and-error problem solving, accidental discoveries, or contacts with other peoples. Steady increases in population through (1) expanded information of natural surroundings, (2) successful expansion of survival techniques by everyday tests, (3) permanent record keeping and permanent organization of knowledge, (4) anticipation of natural events, and (5) social organization and specialization all provide perhaps the best criteria and evidence for the existence of science during this period.

1. Expanded Information about Natural Surrounding

Draining marshes, reclaiming swamps, clearing thickets, eliminating wild animals and reptiles, draining channels, and building dikes around deltas and the mouths of mighty rivers were highly organized social projects that sharpened observation; expanded related knowledge of animal and plant life, of earth, water, sky and weather, of natural processes; and developed techniques and tools for success under extremely adverse circumstances. Such success involved learning and testing by observation and rugged experience. With ensuing increased food supply, specialized labor, discovery of mines, use of metals, navigation and trade, further investigation of surrounding land and waters naturally occurred to supply needs of expanded economic activity. More precise observation of the heavens followed for improved navigation and for anticipation of natural events as well as for religious purposes.

Extended empirical information about natural surroundings during the urban revolutions has already been described. Let us here add or emphasize the following details.

River habitats particularly expanded knowledge of plants, e.g., garlic or herbs as food and medicine, or flax for clothing. Expanded knowledge of animals added to the biological sciences. In about 3100 B.C.E., the Egyptians originated papyrus making from *Cyperus papyrus* reeds that grow along the water, a great step toward written communication and record keeping, basic to science and other forms of culture.

A working knowledge of the physical properties of copper, tin, bronze, silver, gold, iron, and other metals accumulated along with techniques for mining and processing these metals. The metallurgists knew, for example, that heated copper could be melted (at about 2,000°F.) and molded into any shape, and that when cooling, copper hardens to make sharp-edged tools and weapons.

Increased knowledge of mineral properties gave added medicines. For example, malachite (copper carbonate) was used in Egypt to disinfect against eye disease and to protect against sun glare. Interest in bright stones such as opal, agate, lapis lazuli, or turquoise (mined in the Sinai Desert) for trinkets or as mystical amulets resulted in some actual knowledge of semiprecious stones (mineralogy) .

Knowledge of how to harness the wind and improved ship building was acquired in early Egyptian use of sails. Camels had been trained before 3000 B.C.E. by various nomadic peoples. The use of horses to pull Egyptian chariots (ca. 1650 B.C.E.) was an extension of harnessing draft animals. Previous use of wheels for vehicles had provided a revolution in transportation for a number of early cultures.

2. Successful Expansion of Survival Techniques by Everyday Test

The testing of expanding ideas, techniques, tools, discoveries, and resulting products in the urban revolution was by direct trial and error in everyday activities. Testing often led to modification, improvement, and success.

Metal tools and weapons gradually replaced stone. This improvement by successive discoveries modified techniques and facilitated extension of farm lands and settlements. It motivated further discovery, mining, and hauling of

copper, tin, and iron ores (geology and engineering). It involved new knowledge of reducing or smelting ore, a chemical process of heating reddish copper ore in contact with charcoal. In Babylonia it led, as early as 3000 B.C.E.,
to the production of bronze from copper and tin ores—a chemical process.
Around the same time, Babylonia developed a "lost-wax" metal molding
process in which "a wax model of a desired metal object was coated in clay;
the clay was heated to become pottery with the wax running out; and a metal
was poured into the hollow in place of the wax to give a metal casting when
the clay mold was broken."[2] Babylonians shaped weapons, knives, surgical
equipment, scrapers, levers, axes, and other tools from brass produced from
copper and tin, as soon did the Egyptians.

Expanded techniques in the controlled use of fire were shown in both
Babylonia and Egypt in metallurgical and mining developments, in more varied food production and preparation, and in glass making and other innova-
tions of applied chemistry. Early Egyptian colored glass products and mirrors
were famous for their utility and beauty. Babylonians had similar products
from "chemical industries": pottery, glazes and glass, metal products, paints
and dyes, drugs and other remedies, soaps, cosmetics, perfumes, incense,
beers, and wines. Spinning and weaving techniques dramatically improved in
Egypt with the use of flax as early as 3500 B.C.E., resulting in finely woven,
translucent, silk-like linen of the kind seen in royal tombs.

Babylonian tablets show evidence of actual agricultural experimentation
through which to improve the harvesting of dates: farm workers took flowers
of sterile male trees and carried them close to flowers of fertile trees, where
they were left.

Egyptian temples, pyramids, obelisks, and tombs still stand as marvels of
architecture, art, engineering, and astronomy. Babylonian temples and ziggurats were also marvels of mammoth construction.[3] All these represent high
sophistication of engineering with simple, often metal, tools and slave labor.
The oldest Egyptian pyramid, a step pyramid for King Zoser, dates to ca.
3000 B.C.E. and is located near Memphis, the old capital south of Cairo. Constructed entirely of limestone blocks, it rises to a height of 200 feet.

The largest building of ancient times—indeed one of the largest ever constructed—is the pyramid of Cheops at Giza, on the outskirts of Cairo, built ca.
2600 B.C.E. Based on a square, it originally loomed to a height of 480 feet. It
took 2.3 million limstone blocks of up to 15 tons each to produce this pyramid
that was entirely solid except for the funeral chamber and passages to it. Its
precision of construction is still amazing: on lengths of about 775 feet, the difference between the shortest and longest side is only 8 inches! The pyramid's
four faces are aligned precisely north, south, east, and west. Shafts or ducts
were in line with the pole star of the time. Alpha Draconis Lions were considered guardians of holy places. The famous Sphinx made of solid rock with

a 240 foot-long lion's body and a head with royal beard, color, and headdress, is on guard nearby. These two pyramids of Zoser and Cheops are only two of Egypt's eighty major and minor ones.

Another example of highly sophisticated Egyptian engineering are the obelisks. The largest is 137 feet high and weighs 1,168 tons, made of a single block of stone. Quarried in Aswan below the first Nile Cataract, it provides some knowledge of construction and transportation techniques at the time. A stone mass was marked out for detachment in one piece from top layers of granite. The mass was shaped and separated from the surrounding matrix and transported by sleds on moist mud to the Nile. (The Egyptian engineers had found that heavy weights could be easily sledged on ground that had been well moistened into mud.) At the Nile, the obelisk was placed on a ship that carried it to the place of disembarkation from where it was unloaded and further transported by sled to its permanent site. There, with extreme difficulty and care, the obelisk was gradually lifted to an upright position by means of mud brick ramps. One speculation is that "the obelisk may have been pulled up the ramp to a height above its center of gravity. At that point, the mud ramp could have been removed from below until the obelisk settled onto a pedestal with its edge in a pedestal notch leaning against the ramp. From here the obelisk could have been pulled upright."[4] Once the obelisk was in place, intricate and long hieroglyphics were engraved on the hard granite: Obelisks were dedicated to the sun god and included the names of pharaohs, who were considered divinities.

Babylonia conducted huge projects of draining, clearing, and maintaining the Tigris and Euphrates rivers for agricultural purposes; built canal networks for irrigation, transportation, and communication; constructed huge temples as religious and administrative centers; and used the attached ziggurat towers as massive astronomical observatories. In Babylonia as in Egypt, applied mechanical principles in irrigation and canal building as well as in handling huge stone structures were highly, even amazingly, successful with basically simple devices. Both Egyptian and Babylonian farmers used simple "shadoofs" for leverage in raising water from canals to land. Highly successful Egyptian mummification techniques are still marvels of functional chemistry that included manufacture of many other metal, mineral, glass, drug, paint and dye, cement and other chemical products by specialized workers or artisans.

Egyptian surgeons were advanced in their day for techniques even for plastic surgery. The nature of surgery itself involved empirical knowledge and direct treatment of body wounds and injuries. Surgical experience thus left less room for care based upon the mystical or supernatural as compared to general medical practice. The famous Egyptian Smith (medical) Papyrus (ca. 1800 B.C.E.) reveals organized knowledge of anatomy and physiology, indicating approaches to surgical practice and specific case illustrations. The equally famous Ebers Papyrus (ca. 3000 B.C.E.) contains 877 medical recipes in divisions of internal medicine and surgery that included diseases of the eye, nose, throat, skin, and limbs as well as women's and surgical diseases.

Babylonian surgery was also quite empirical, and was governed by the impressive legal code of Hammurabi (ca. 2000 B.C.E.), about two thousand years before the famous Roman legal codes. As compared to surgery by trained empirical artisans, however, Babylonian medicine was theocratic and in the hands of priest-physicians. Diseases were of divine, demonic, human "evil eye," or animal magnetism" origin. Remedies, therefore, by Babylonian physicians, generally included incantations, prayers, sacrifices, other appeasements, magical rites, even exorcism. There was also some use of herbs and drugs. Placebo effects most likely existed in the majority of the above "medical" practices.

3. Permanent Record Keeping and Knowledge Organization

In both Babylonia and Egypt, written language and mathematics were invented; information was recorded and kept; libraries were built; and special scribes (generally priests) were trained. That is, written records, whether on papyrus scrolls or clay tablets, were kept for permanent accumulation, organization, and transmission of information. Such practices are basic to science as well as to general culture. More specifically, Babylonian and Egyptian administrator-priests served as trained record keepers, tax collectors, sky watchers, and other civil service personnel. They developed standard systems of weights and measures, accounting systems and mathematics for land measurement, tax collecting, mammoth building construction, huge irrigation projects, sky mapping, trading, and the like.

In all cultures a prerequisite to permanent records of scientific and other data are language and mathematical symbols. Egyptian pictograms or hieroglyphics representing ideas or images on stone are well known today, particularly in connection with excavated royal tombs and walls. Papyrus provided a new, easier, cursive running script called "hieratic" on scrolls as compared to hieroglyphics, on stone. The Egyptians originally, in connection with hieroglyphics, also developed auxilary, supplementary alphabetic signs. It was the Phoenicians, early natives of present-day Lebanon, however, who first gave to civilization a language fully based on an alphabet.

Mathematics is also a symbolic language and logic recognized quite early in Egypt and Babylonia. For example, the Egyptian Rhind Papyrus (ca. 1600 B.C.E.) reveals much about the development and applications of mathematics. This development, it seems, arose primarily from economic and other cultural needs. Examples abound in land masurements for refixing boundaries after floods; in length, area, volume, and weight measurements for trade; in other accounting applications; or as aids in the construction of huge temples, tombs, pyramids, or granaries. Unique arithmetic techniques for handling fractions are also illustrated in the Rhind Papyrus.

The later Greek Pythagorean theorem took a preliminary arithmetic form of $3^2 + 4^2 = 5^2$ in Egypt. The depiction is entirely functional. The 3, 4, and 5 were marked off on ropes to give right-angled corners in building construction. But nowhere to be found in Egyptian records was any generalized algebraic formulation of the Pythagorean theorem ($a^2 + b^2 = c^2$) stating that in *any* right triangle, the sum of the squares of the two sides equals the square of the hypotenuse. There were, however, elemental algebraic and geometric formulations of areas and volumes as well as for areas of triangles or volumes of pyramids, even of truncated pyramids. And also existing with direct empirical need were other mathematical relationships among lengths, areas, volumes, and weights.

Babylonian mathematics in general was even more advanced than in Egypt. They originated a *position* concept of numbers with a 60 base preceding our present base of 10 (that is, our metric arrangement of 1, 10, 100, 1,000, etc.). It was the Babylonians, however, who originated the metric (decimal system) idea by also using 10 as a factor of 60. A weakness in their position concept of numbers was the absence of a symbolic zero (0). Our present sexagesimal division of time (60 seconds = 1 minute; 60 minutes = 1 hour) goes back to the Babylonians. So does our sexagesimal (protractor) division of angular space into 360 degrees in a circle, 60 minutes in a degree, and 60 seconds in a minute.

The Egyptians and Babylonians developed algebra (generalized arithmetic) from agricultural, trade, land, tax, and other business accounting needs. Their algebra included linear, simultaneous, quadratic, and cubic equations, revived by later Arabs and unappreciated in the West until the time of Descartes in the seventeenth century. Even Galileo (1564–1642) was unfamiliar with algebra as such. Babylonian beginnings in geometry were comparable to those in Egypt, including areas and volumes of geometric figures, Pythagorean theorem relationships (but in algebraic form), and the value of *pi*.

Standard weights and measures arose in trade and other economic relations in the urban revolutions of Egypt and Babylonia as well as India and China. Standard units of length were often based upon parts of the body, as the finger or foot. For example, in Babylonia, a "span" was a width of 15 fingers, with 2 spans equal to one cubit—again based on their sexagesimal system. Our standardized "foot" of length today is a cultural heritage of ancient peoples. Volume was often expressed by the sackful, and weight by counterbalancing standardized pieces of metal or stone. Among measuring rods for accurate azimuth angle measurements in astronomy, Babylonians and Egyptians used plumb lines and forked rods. Basic standard units of space, time, and weight (or eventually mass) were indispensable in any development of theoretical, conceptual systems in science or mathematics.

Time-keeping devices, such as waterclocks and sundials, were also developed in Egypt and Babylonia. The sundial, with its shadow-length measurement, was used as a daily and yearly timepiece as well as a direction finder. Waterclocks enabled night timekeeping in astronomy, for example. Sundials were day timekeepers. The Egyptians varied the length of their hours daily so as always to have twelve hours between sunrise and sunset. For accuracy in astronomical computations, the Babylonians were the first to standardize their twenty-four hours to be equal throughout the day and year.

As early as 3000 B.C.E., the Babylonians developed a monthly calendar averaging $29^{1}/_{2}$ days because that is the time interval from new moon to new moon. The word "month" is really a contraction of "moonth." This totaled 354 days in twelve months. An intercalated thirteenth month was added from time to time to reconcile the calendar with the seasons. The Babylonians originated the idea of a seven-day week but with the first of each month as the first day of a week. (That meant an incomplete week at the end of each month.) When the Hebrews developed agriculture, they adjusted their lunar calendar to the Babylonian, but with equal, continuous weeks for regularity of sabbaths and religious calculations.

The Egyptians had a 10-day week called a "decan." A wide belt of constellations had been divided into 36 parts, 10 degrees each. Each group of stars would rise with the sun on succeeding 10-day periods. Much more important in Egyptian contributions to astronomy was establishment of the solar calendar, a conceptual model originating from a repeated natural event: the yearly overflow of the Nile. The flooding coincided with the regular heliacal rising of the star Sirius. ("Regular heliacal rising" refers to the specific observations that any given star rising with the sun does so again after a certain interval of time.) The observed interval of time between the heliacal risings of a star is so exact, regular, and general that the Egyptians made it the basis for the year as a unit of time. The flooding of the Nile had enough economic importance to start the Sothic or astronomical year with the heliacal rising of Sirius (Sothic) associated with the flooding.

Egyptian astronomer-priests divided the year into twelve months with three decans a month (30 days each) for a total of 360 days. This required a year-end holiday season of five days for 365 days. Egyptian astronomers noted, however, that the heliacal rising of Sirius fell behind the approximate 365 day yearly calendar by about one-fourth of a day. This meant that the year was actually $365^{1}/_{4}$ days and that each calendar day had four chances to be the day for the heliacal rising of Sirius, i.e., to be New Year's Day. This created a Sothic cycle of 1,460 (4 x 365) days before Sirius (Sothis) would again rise with the sun on the *original New Year's Day.*

In sum, the Egyptian solar calendar was scientifically innovative and unique. The flooding of the Nile was not a caprice of an Egyptian god but a predictable yearly event associated with the heliacal rising of Sirius. A calendar associated periodic floods with regular motions of sun and stars. Culminating from systematic record-keeping of natural events, it was a scientific conceptual model that helped "call the world to order" through anticipation and organization of events.

4. Prediction of Natural Events

Babylonian observation of the heavens was exceptionally accurate, mathematical, and well organized for its time. Most likely, astrology originated with the Babylonians. Religious and business purposes behind the astrology demanded careful, precise observation, records, and organization. Ziggurats with their high brick towers afforded observation posts of the heavens. Babylonians used their 60-base number system and their geometric insights to project a zodiacal belt with twelve signs or constellations of stars upon the heavens. Each constellation occupied a space of 30° for a complete circle of 360°. This belt was about 8° on each side of the ecliptic, the observed path of the sun in eastward drift among the zodiacal constellations. We have here the earliest organized records in the use of degrees and regular measurements in the sexagesimal system used in protractors today. We also have in the zodiacal belt and its details the projection of a scientific model for prediction of celestial events. All then known planets as well as the moon and the sun were found always wandering within the belt. Accurate continuous records were kept, at first, particularly of Venus, the sun, and the moon. These records show definite regularities in the motion of these celestial bodies. Soon other records revealed similar regularities for other wanderers of the sky. Based on charts, the Babylonians could accurately predict the first and last appearances of Venus as both a morning and an evening star. There was also prediction as to length of future disappearances of Venus. Based on charts, the daily, monthly, or yearly motions of the sun, moon, and planets in the zodiacal belt of stars thus became predictable. And prediction is at the heart of science. Eventually the Babylonians could also anticipate eclipses of the moon and sun. At the time, predictability was a powerful persuasive weapon in the hands of Babylonian priests reading astrological messages in the sky inspired by the gods. But for our purposes here, we see Babylonian zodiacal predictions of planetary positions as a basic scientific development resulting from permanent record keeping, organization of knowledge, and conceptual model projections even though motivated by religious, politico-economic, and astrological objectives.

Thus, the Egyptians had their solar calendars and the Babylonians their zodiacal belt as conceptual models for astute observations. These models are scientific and sophisticated enough to survive even today in modified form. The significance of scientific conceptual models in human interaction with nature will be further emphasized in later chapters.

SUMMARY

Science began when men and women first learned by experience how to survive. It developed in prehistory and in the agricultural and urban revolutions with expanding survival skills, observations, and insights gained through experience. Science and its human interactive character can be seen in (1) the unquestionable expansion of techniques, ideas, and products which humans with tools used as extended muscles and adapted to their surroundings; (2) the actual *testing* of these techniques, products, and ideas by direct survival use and by pragmatic *rejection or modification* of what did not work; (3) the *transformation of surroundings by large-scale social organization* for better survival (e.g., urban revolutions of Egypt along the Nile and of Babylonia in the Tigris-Euphrates plains; (4) development of specialized skills and techniques within such diverse activities as mining, metallurgy, metal smithing, gigantic public building construction, intricate canal and irrigation systems, and expandling arts and crafts; (5) the development of language, writing materials, mathematics and record-keeping systems; (6) organized observation of celestial motions; (7) anticipation of natural events as the yearly flooding of the Nile or lunar eclipses; (8) eventual development of scientific conceptual models as solar calendars or zodiacal belts for organization and prediction of natural events or for organization of human activities.

In their interaction with nature through their senses, minds, imaginations, and muscles, humans were successfully adjusting to surroundings and yet effectively changing their surroundings. Thus, the observation, systematizing, description, prediction, and testing of natural events were all in place. Conceptual models, too, as well as feedback modification from experience with nature. That is, in human interaction with surroundings, techniques were mounting for coping and for learning from experience. That was science. There was a world of things and a related world of ideas. And there were observed patterns of motion in the vast heavens. These, too, are typical of science today. The world of things involved realities with which to cope. The world of ideas at first explained the world of events through capricious spirits and gods that needed to be appeased for favorable events. Elementary anticipation and causality were there is supernatural form. Recorded regularities of the motion of sun, moon, and planets against a Babylonian zodiacal belt enabled production of future relative positions of these bodies. Such predictions were at the heart of a scientific astronomy even though the predictions have been identified with the use of astrology for divine portents by Babylonian priests. The Babylonian civilization was eclipsed when it fell to the Persians in the seventh century B.C.E. That was the same century in which Ionian Greek philosophers conceptually reduced all objects to basic elements in order to explain transformations of matter and thereby explain natural processes through the workings of nature itself. These were followed by the Greek atomic models of Leucippus and Democritus. Explanation through natural causes had arrived, soon to show itself in an expanded, sophisticated Greek astronomy, geometrically shaped.

NOTES

1. *National Geographic,* May 1972, p. 63.
2. V. Gordon Childe, *Man Makes Himself* (New York: Mentor, 1963), ch. 7.
3. Ibid.
4. Ibid., ch. 8.

SUGGESTIONS FOR FURTHER READING

Bernal, J. D. *Science in History.* Cambridge Mass.: MIT Press, 1977, vol. 1, part 1: "Emergence of Science," and part 2: "Science in the Ancient World."
Childe, V. Gordon. *Man Makes Himself.* New York: Mentor, 1963.
———. *What Happened in History.* Baltimore, Md.: Penguin, 1957.
Hogben, Lancelot. *Science for the Citizen.* New York: Norton, 1951, ch. 1: "Pole Star and Pyramid—The Coming of the Calendar."
Neugebauer, O. *The Exact Sciences in Antiquity.* New York: Harper, 1962, chs. 1–5.
Pannekoek, A. *A History of Astronomy.* New York: Outerscience, 1961, chs. 1–8.
Sarton, George. *A History of Science.* New York: Wiley, 1964, "Oriental Origins," chs. 1–4.
Taylor, Lloyd W. *Physics, the Pioneer Science.* New York: Dover, 1941, vol. 1.

3

Natural Causes:
Ancient Atoms, Elements, and Transformations of Matter

Atomism has proved the power of the intellectual imagination to identify aspects of an objective truth deeply rooted in the nature of things. Hidden in the history of atomism . . . there must be still concealed a trustworthy foundation for the human intellect. . . . No one is so brilliant that he can afford to neglect what history can teach him.

L. L. White, Panel Discussion, San Francisco State University, 1961

ANCIENT GREEK IDEAS OF MATTER

Atomism originated among those ancient Greeks who went in search of a basic alphabet of nature. Atoms were an idea long before they became a fact of technical power that will either solve basic human problems or destroy civilization.

Have you ever considered that the countless ideas in thousands of volumes in a library are arrangements of but twenty-six letters of the alphabet, ten numerals, and a few punctuation marks? And have you ever thought about how all the rich profusion of objects around us are formed of a basic alphabet of nature itself in various combinations? Or wondered what the character of nature's "building blocks" would be? The Greeks *did* wonder about such things. They were intrigued by the diverse, changing, temporary character of objects around them. But they were confident that the universe is a unity and they looked for something unifying and eternal in the variety and flux of things. In the search for order, Alcymaeon of Croton (sixth century B.C.E.) likened the universe to the human body, with all parts correlated to the whole. This analogy gave astrologers something to grasp. The relationships of the stars to man and his activities became all the more apparent as the working and relationships of parts in a whole. Plato, in his *Timaeus*, reinforced this in an animistic analogy. The universe is a living organism of body, mind, and soul.

Any particular material object has an importance that is secondary, changing, and temporary within the underlying, permanent reality of the whole—just as the cells and tissues of each one of us undergo wear, tear, and replacement in the rare fundamental reality of our complete self. Common origins of poetry, religion, philosophy, art, and science in man's search for order (and survival) are revealed by such analogies. The Ionian Greeks were the first materialists. They reduced all objects to elemental components in order to explain the variety and transformations in matter. Their idea was to look into matter itself to understand matter. Successors attempted to reduce objects to minute particles. When Leucippus in the fifth century B.C.E. pictured these unseen particles as hard, indivisible, unchanging, and eternal, atomism was born. "Atom" means "indivisible" in Greek. Things change, but irreducible units composing them can be permanent.

Plato, Aristotle, and the Roman philosophers Seneca and Lucretius describe Greek ideas of matter in detail; yet these men based their own ideas upon predecessors whom they acknowledged. Lucretius, in his *The Nature of the Universe,* provides a vivid exposition in poetic form of Epicurus's atomism, which, in turn, was originated by Leucippus and further developed by Democritus. Plato, in his *Timaeus,* falls back upon the Pythagorean idealization of number and geometric forms as the essence of things. Aristotle, in several works, elaborates the idea of the four elements (air, earth, water, and fire), which he attributes to Empedocles. Table 3.1 summarizes Greek ideas of matter and provides insights into the theoretical foundations of chemistry. Conceptual emphasis is on the nature of elements and on transformation of matter.

One Basic Element

The one-element Ionians (Table 3.1) were the first systematic natural philosophers in Greece. They had a naturalistic, materialistic bent: the Ionians sought causes and explanations in terms of the eternal working of things themselves rather than in any divine, mythological, or supernatural intervention. Looking for a single basic reality, the Ionians believed that all things have their origin in a single knowable element: water, air, fire, or some indeterminate, nebulous substance.

To Thales (born ca. 624 B.C.E.), the originator of the Ionian school, the primary element was water. As Seneca states it in his *Physical Sciences,* "Water is, according to Thales, the most powerful of the elements. He thinks it was the first of them, and that all the others sprang from it. Also, the whole earth is upborne by water, and floats just like a boat." Why did Thales select water as "primary"? Any answer to that question is speculative. We can merely guess that he was impressed by the widespread prevalence of water, whether on the earth's surface—three-fourths of which is water—in the air, in plants, in animals, in almost everything everywhere. Perhaps he was impressed by the falling of rain, by the melting of ice or snow, by the weathering of soil and rock. In the freezing of water he saw the formation of solid matter. In boiling water he witnessed the formation of vapor, or what he considered air. In this process, he also may have noticed the slight residue of solid matter that the water left behind. Nowhere does Thales account for fire, however, or give specifics about how material transformations take place. Thales and his successors, however, showed awareness of physiological changes in which air, food, and water taken in by plants and animals are altered when eliminated.

Anaximander's (ca. 611–547 B.C.E.) primary element is an eternal, endless substance that Aristotle referred to as a kind of primal "chaos" or nebula. Evidently, Anaximander was not satisfied that water or, for that matter, fire, air, or earth is primary. There has to be something common to all of them, since in freezing, boiling, or other processes, one element seems to be transformed into another. In a world of changing substances and one in which Anaximander recognized a strife of opposites (water versus fire or cold versus heat), the primary element has to be permanent, eternal, and boundless. Otherwise, the universe would run down in the continual waste and destruction of ordinary things we know. Anaximander visualized all substances as originating from an eternal primary mass through separation of opposite qualities, such as hot and cold, wet and dry, to form water, fire, air, earth and a world of things. Then from the "strife" of opposite substances previously formed comes the continual formation of fresh or new substances. Yet Anaximander believed that everything would eventually settle back and disappear into the general "indeterminate" mass—just as scientists say today that the universe is running down or leveling off because of an irreversible dissipation of heat in *natural processes* (the second law of thermodynamics).

To Anaximenes (sixth century B.C.E.), all things originated from air. Here again, we can only speculate about why he believed this. He may have been influenced by the fact that air is everywhere around us, even within us, and that in air all kinds of things are observed: solid objects such as dust, snow, or the moon; liquids, such as clouds or rain; or vapors, as indicated by odors. We do know, however, that Anaximenes ingeniously explained the transformation of air into other substances by condensation and rarefaction. For example, air rarefied becomes fire; air somewhat condensed becomes wind; more condensed, water; still more condensed, earth; still further condensed, stones; and so forth. Thus, all substances vary quantitatively only; the only differences are ones of density.

Heraclitus (ca. 540–475 B.C.E.), like Anaximander, was primarily impressed by two things in the world around him: first, the continual state of change and flux he saw everywhere; second, the ever-present struggle of opposites. With everything temporary, limited, changing, and in strife, there

Table 3.1. Ancient Greek Ideas of Matter

	Nature of the Elements	Transformations of Matter
THE ONE-ELEMENT IONIANS		
Thales (ca. 624 B.C.E.)	Water.	Bodies of plants and animals ⇄ Air ⇄ Water ⇄ Earth (cyclic)
Anaximander (ca. 611–547 B.C.E.)	An unlimited, eternal, indeterminate substance.	Eternal primary substance → Various substances, e.g., water; Eternal primary substance; Opposite substances, e.g., fire. Transformation through separation and "strife" of opposites, moist, dry, etc.
Anaximenes (6th century B.C.E.)	Air.	Stones, etc. → Earth → Water → Air → Fire. When rarefied. When condensed. Differences among all substances merely quantitative.
Heraclitus of Ephesus (ca. 540–475 B.C.E.)	Ethereal fire of everlasting, driving character.	Primordial Fire → Various substances, e.g., wood → Primordial Fire; Other substances, e.g., smoke
THE FOUR-ELEMENT PHILOSOPHERS		
Empedocles ca. 490–435 B.C.E.	Fire, air, water, and earth. Empedocles proved that air is a substance by showing that water can enter a vessel only as air escapes.	1. Four elements not interconvertible. 2. All substances are combinations of four elements caused by attraction (love) or repulsion (strife).
Pythagoreans From 5th century B.C.E. on Philolaus (480 B.C.E.–?)	1. Numbers are elemental realities of everything 2. Fire, air, earth, and water are built up out of regular geometric solids: a tetrahedron, octahedron, icosahedron, and cube, respectively. 3. The geometric solid dodecahedron is a fifth element.	The multiplicity of things is understandable in the relationships and transformability of numbers and of geometric forms.

Plato (427–347 B.C.E.)	1. Ideal types or qualities of things are the permanent realities giving temporary form to substances as we know them. 2. In his *Timaeus*, Plato also subscribed to the Pythagorean version of the four elements.	
Aristotle (384–322 B.C.E.)	1. Primary essence of matter: four qualities of *hot, cold, dry*, and *moist*. 2. Fire, earth, water, and air, derived by combinations in pairs of the above four qualities as: Fire — dry — Earth — cold — Water — moist — Air — hot (Fire) E.g., fire, hot and dry; earth, dry and cold, etc. 3. A fifth ethereal element of eternal, almost divine quality beyond the human sphere.	1. Fire, earth, water, and air *are* interconvertible. 2. All substances consist of the four elements variously combined in different amounts of the four qualities in pairs. 3. The fifth ethereal element is not transformable.
THE IONIAN OF COUNTLESS ELEMENTS Anaxagoras (510–428 B.C.E.)	*"Seeds"* Each substance is infinitely divisible into "seeds" of itself.	No transformation: There is a "portion of everything in everything."
THE ATOMISTS Leucippus (ca. 500 B.C.E.–?) Democritus (ca. 460–370 B.C.E.) Epicurus (ca. 342–270 B.C.E.) Lucretius (ca. 98–55 B.C.E.)	*Atoms* 1. Eternal. 2. Indestructible. 3. Minute. 4. Solid. 5. Indivisible. 6. All identical in substance. 7. But variable in size, shape, and weight. 8. Constantly moving.	Individual atoms and void ⇄ (separations / collisions) Groups of atoms Substances, once formed, vary according to size, shape, weight, and grouping of atoms.

had to be a single, underlying reality, permanent and driving in character. Fire was such a dynamic force-substance. Heraclitus visualized a primordial ethereal fire, ever burning, consuming fuel and releasing smoke, taking some substances into itself and, on the other hand, giving off other substances—as against the indeterminate nebulous substance of Anaximander. Like the latter, Heraclitus also believed that just as everything originally arose from his primordial substance, fire, everything would eventually settle back into it. And Seneca concurs in the *Physical Sciences*: "We Stoics also say that it is fire that lays hold upon the world and changes all things into its own nature." So fire is the beginning and the end.

The Four Elements

Both Aristotle and Lucretius attributed the concept of four elements (chart 3.l), earth, water, air, and fire, to Empedocles (ca. 490–435 B.C.E.). Pythagorean contemporaries also used the idea. The Pythagoreans, however, emphasized the primacy of geometric forms as the essence of the four elements rather than the primacy of the four elements themselves. Most likely they fitted Empedocles' idea into their own.

The Egyptians had long recognized four elements, and imbued them with male or female qualities. For example, earth is male when it has the form of boulders and crags; female when it is cultivable land. When air is windy it is male; when cloudy or sluggish, female. John Read, in his *Prelude to Chemistry,* claims that in both India and Egypt, the four elements, had been a fundamental idea since about 1500 B.C.E., and that the Chinese, as early as the twelfth century B.C.E., had five elements, wood in addition to the above four together with five virtues, tastes, colors, tones, and seasons. We may merely speculate as to whether the number 5 had significance here to the Chinese, just as the number 4 did to the Pythagoreans in their original designation of four basic geometric forms and four elements of matter.

In any case, Empedocles was the first Greek natural philosopher to claim there is not one but four basic elements. He was aware that Ionian philosophers had chosen water, air, and fire singly as primary elements. Empedocles could not see that any one of these alone would adequately explain the tremendous varieties and complexities of substances. He believed that Heraclitus's fire had to have elements of water already within itself to give rise to its opposite, water. Empedocles believed that he had solved the problem by considering all four instead of one of the elements as primary. That could enable him to consider all four "indestructible, unchangeable, and permanent." That would also enable him to use all or various proportions together to explain the large diversity of objects, just as present-day chemistry uses its more than one hundred elements. To Empedocles, elements do not transform one *into* the other;

they *combine* with one another to form the world of things. That is what makes them basic. He explained, for example, that bones consist of a proportion of 2:1:1 of fire, earth, and water as against a 1:1:1 ratio of these elements for flesh and blood. A crude chemistry, indeed, but one that seeks answers within things themselves. In fact, in his emphasis on different fixed ratios of elements in different substances, Empedocles was over two thousand years ahead of his time!

Operating universal "forces" of Love and Strife, attraction and repulsion, were behind combinations and recombinations of Empedocles' elements. Strife was the struggle of opposites emphasized by Anaximander and Heraclitus, to which Empedocles added the attraction principle of Love. The elements and all substances were attracted into and out of combination with one another under the influence of these universal "forces."

Empedocles also proved the material of the element air by demonstrating that more water enters an inverted container as air escapes (Fig. 3.1). No longer could vacuum, air, and space be synonymous.

To the Pythagoreans, numbers were "the reality of everything." Numbers were identified with space and matter in the following way: The number 1 is a point, or a unit, for indicating position and size. Arrangements of such points give geometrical figures: Two points determine a line; three points, a surface; and four, a solid (Fig. 3.2). The number 4, representing solids, applied to Empedocles' four elements.

Aristotle's physical ideas of matter may be summarized as follows:

1. His ideas were an elaboration of Empedocles' four elements, already described.

2. He rejected Empedocles' idea that the elements of earth, water, air, and fire cannot be transformed into one another. He did so by making Anaximander's two pairs of opposite qualities (hot and cold, dry and moist) primary to the four elements, and in combinations the essence of them. For example, fire is formed from the qualities of *hot* and *dry,* earth from *dry* and *cold,* water from *cold* and *wet,* and air from *wet* and *hot* (Table 3.1, Aristotle). To transform fire into earth, all that is necessary is to replace *hot* with its opposite, *cold*; or to obtain earth from water, supplant *wet* with its opposite, *dry.* However, not only do the elements transform one into another through these qualities, but in various combinations of the four elements with pairs of the four qualities, all worldly, sublunar substances are formed. Only two of the four qualities can be used at any time with the four elements, "for it is impossible for the same thing to be both hot and cold, or moist and dry."

3. Aristotle introduced a fifth element. This element, however, unlike the first four, cannot be "generated, corrupted, or transformed." This pure, eternal,

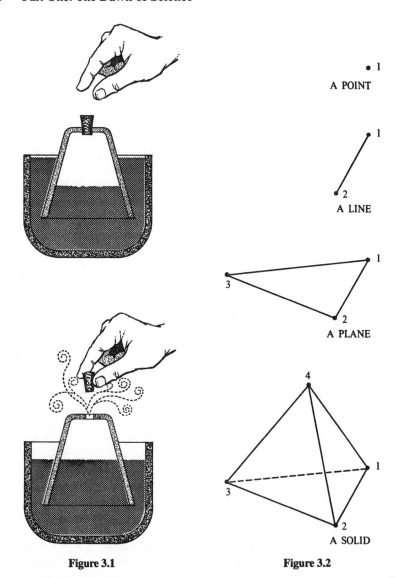

Figure 3.1	**Figure 3.2**
Empedocles' air experiment. When the stopper is removed, the water in the inverted glass rises. Conclusion: Escaping air is invisible matter that occupied space taken by the water.	Numbers are to be identified with points that determine lines, planes, and solids (Pythagoreans).

ethereal substance makes up the heavens and all their unchangeable objects. Beneath the lunar sphere, the many products of the terrestrial four elements are ceaselessly transformed. While the four elements ever seek their various places, the fifth reveals its divine, ethereal quality in perfect, circular motion. Aristotle points out, however, that although all substances of the four elements are individually "generated, corrupted, and transformed, the universe as a whole is ungenerated and indestructible." Thus, in Aristotle also we find a conservation-of-matter concept.

Seeds of Matter

Anaxagoras in his "seeds" idea (Table 3.1) was unwilling to submerge the tremendous varieties in things into any common denominator, whether it be the concrete water of Thales, the air of Anaximenes, or the intangible ethereal fire of Heraclitus. He preferred to accept the immediate diversity of things as they are. In Anaxagoras's reasoning, every object is infinitely divisible. Every tree can be a sliver; every sliver, a speck; and every speck halved without end. No matter how far this process goes, what is left would have characteristics of the original substance. That is, the broken-down speck has the characteristics of wood in the same sense that today a molecule has basic properties of the substance it composes. We, however, stop at the atom or molecule for similar properties, while, as Lucretius pointed out two thousand years ago, Anaxagoras did not stop at all. Anaxagoras was intent, however, in making his basic point that every item is made up of unique miniatures of itself.

That brings us to the very interesting question of the transformation of matter. If all substances are derived from unique seeds of themselves, how can one substance develop into another? How does the potato you eat become body fat? Anaxagoras's answer was that every substance down to the utmost minute fragment contains a portion of everything else within itself. That it is one substance rather than another is only because of the relatively greater amount of itself that it contains. If you eat potatoes and get fat, it is because potatoes provide fat as well as starch and water. Your body stores the fat. Such a conception of a minute substance containing something of everything within itself seems fantastic until we realize that today, in referring, for example, to the "half-life" of a meson, we are talking in terms of billionths of a second. Can we really conceive of how small a billionth of a second is? Because something is inconceivable to us does not necessarily make it incorrect. Our own present limitations are involved.

The contribution of Anaxagoras's concept of "seeds" was its refinement, its idea of taking substances as they are and breaking them down minutely in order to know more about them. No scientist today will argue against analyzing things. Anaxagoras's seeds, however, were too complex. Anaxagoras was passing on the complexities of a large-scale object to unseen miniatures of itself. And since these miniatures were infinitely divisible, a fundamental unit was lacking.

Indivisible Atoms

The Leucippus-Democritus atom (Table 3.1) combined features of the Ionian single element, Anaxagoras's "seed," and Empedocles' four elements and yet was an improvement over all of them. Let us see why. There was the great benefit of simplicity in the single-element idea of the Ionians, but the idea was too simple. A single element, water or air, could not account for the endless number of different things in the world without changing its quality or losing its identity as an element. Empedocles attempted to give flexibility by using all four elements as basic. What is more, he established elements as permanent, indestructible, and unchangeable; they did not lose their identity; yet they could not get far enough in a complex world. As Lucretius argues in *The Nature of the Universe,* if these four elements are combined:

> In the wild congress of this varied heap
> Each thing its proper nature will display,
> And air will palpably be seen mixed up
> With earth together, unquenched heat with water,
> But primal germs in bringing things to birth
> Must have a latent unseen quality.

In other words, the above four elements would form mixtures of themselves only, not new substances, unless they contained something more primary than themselves. Besides, said Lucretius, earth, air, water, and fire are too "soft" to be elements. In the actual world they are destructible, changeable, variable.

Let us see how Leucippus's atoms combined to good advantage the simplicity of the Ionians; the indestructible, unchanging, and indivisible unit idea of Empedocles; and Anaxagoras's analysis of individual objects. "Atom," as we have said, means "not divisible" in Greek. That term was intentionally chosen by Democritus to emphasize a particle so small that it could no longer be divided. To Leucippus and Democritus, originally and basically the universe consisted entirely of "atoms" and a "void" in which atoms moved. Thus, the atom in its "solid singleness," as Lucretius expressed it, became a particle common to all substances. The atom is eternal; there was no need to look for anything primary or Primary in it (definite materialism here). Being eternal, the atom is indestructible and unchanging.

Atoms are all of the same substance; but by varying in size and shape, they can be used to explain the large variety of objects they compose. Water, for example, "as a liquid has its atoms smooth and round, gliding over each other, while atoms of iron are hard and rough." Lucretius believed that substances like air, earth, and water are soft because there is "void" (*vacuum*) between the atoms composing them. On the other hand, hard substances like iron are so only because the iron atoms, themselves hard and solid, are more concentrated.

However, differences in size, shape, and weight were only part of the explanation of diversity. All atoms, like today's molecules, are continually in motion and remain that way naturally unless stopped—advance shades of Galileo's inertia! Democritus visualized these atoms moving in all directions in a whirl. Atoms group, and a world of substances results. In the general whirl, atoms of like size, shape, and weight are sorted out and thrown together to form large concentrations of a given substance, just as by centrifugal force we separate skimmed milk from cream.

Lucretius, a student of Democritus and Epicurus, visualized atoms as moving in orderly, parallel fashion rather than in whirls, but with "a little swerve at times." These "swerves," of course, result in collisions, substances, and worlds, although not all collisions are fruitful. "All visible objects are compounds of different kinds of atoms," explained Lucretius. Atoms of different sizes and shapes may be thrown together, separate, and regroup variously. Individual atoms, solid, eternal, and indestructible, always maintain their identity in uniting or separating. it is the union and separation of atoms that is temporary and that results in the transformation of objects. The total number of atoms remains the same as atoms group and separate in transformations of matter. The result is the principle of conservation of matter: it is neither created nor destroyed but transformed.

Individual atoms have no color, sound, odor, or heat. These are all secondary characteristics of a group of atoms that change as the group changes. These characteristics are applicable to perishable objects and not to the eternal individual atoms composing objects.

After the atomists, the ideas of Plato and Aristotle on the nature of matter may seem almost an anticlimax. Without, however, the possibility of experimentation or the refinements of measurement, the Greek atomic theory, as it turned out, did not have a chance against the authority of Plato and particularly that of Aristotle. Besides, the four-element theory had evidence of its own, as illustrated by a lit candle (Fig. 3.3) . The dancing flame apparently is the element *fire* escaping from the candle as the wax burns down. Warm *air* also can be felt rising above the visible flame. Place a cold piece of glass above the flame, and small drops of *water* can be found forming on the glass. After the candle has burned out, remaining ash and soot are the *earth* element left behind. In the burning process, the elements fire, air, water, and earth have been freed from their union in the candle and return to their natural places in the universe. In a reverse order, living creatures breathe in air, drink water, eat

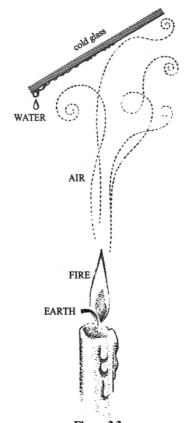

Figure 3.3

Four-element theory "evidence": earth, fire, air, and water replace a burning candle.

food, and have warm bodies. Heated food contains "fire." The four elements are compounded within.

We, of course, no longer accept air, water, earth, and fire as basic elements. We now have well over one hundred of our own, and the number continues to mount. But the concept "element" was a milestone in the history of ideas. And the attempt to understand objects in terms of the workings of their respective parts was inspired. In general, ideas of any given period can best be understood against the framework of their own times. Each individual—even if he be an Aristotle, a Newton, or an Einstein—rests upon the accumulated knowledge of his day, and if he (or she) is outstanding, he takes a step forward from there. We have seen that whether it was the Ionians who emphasized a single element, Anaxogoras and his "seeds," the atomists, Empedocles and the four elements, the Pythagoreans, Plato or Aristotle, one borrowed from the other. Even when the thinking took different lines or directions, as with the atomists and Aristotle, they could still be traced back to common roots. The atomists developed from and in reaction to ideas of single elements, and so did Empedocles and Aristotle. And the thinking of all was conditioned and limited by the particular accumulation of knowledge of the time. The atomists were the closest to present-day ideas of matter. But experimental techniques and knowledge had not yet reached the point where they could prevent the atomists from being eclipsed by Aristotle—and experimental techniques and knowledge are social accumulations. Who knows what ideas, eclipsed today, will emerge illuminated tomorrow?

ATOMS, ELEMENTS, AND THE SCIENTIFIC REVOLUTION

Again, four elements were too simple an idea. As we shall see in later chapters, men search for order through a few unifying principles. But nature has a way of seeping through categories established for it. Meanwhile, men gain detailed knowledge of their surroundings and develop more sophisticated concepts, techniques, and unifying principles. Experimentation, which became respectable in the Scientific Revolution, eventually showed that earth, water, air, and fire can be resolved into simpler substances and that the Aristotelian elements are not elements after all. Or that they are more appropriate as states of matter that apply to every substance. That is, substances called "earths" are solids; "water," liquids; "air" and "fire," gases.

Francis Bacon was the seventeenth-century trumpeter of the Scientific Revolution, which Copernicus sparked unknowingly. Bacon was aware of the work of Galileo, Kepler, Gilbert, and other contemporary experimentalists. In place of science by authority, he appealed to experimentation. He expressed respect for men of antiquity who "used to deliver the knowledge which the mind of men of antiquity who "used to deliver the knowledge which the mind of men had gathered in observation . . . [and who invited] men both to ponder that which was invented and to add and supply further." But Bacon, in his *Novum Organum* (1621), sharply criticized his own contemporaries for whom "sciences are delivered to be believed and accepted, and not be examined and further discovered." "Go to the horse's mouth and count teeth rather than quote authorities," Bacon challenged. Wilhelm Conrad Roentgen, the discoverer of X-rays three centuries later, had the right idea when he said: "What did I think? I didn't think. I began to experiment." Rather than force facts into preconceived ideas, go to nature, get the data, and then form conclusions. Bacon was a crusader for a direct fact-finding, inductive approach to nature.

In any case, ancient Greek concepts of matter were highly significant historically: They explained early on natural events in terms of the workings of nature itself rather than through supernatural agencies or intervention. Behind natural events are natural causes. Matter was reduced to unseen atoms which were to be used to explain natural phenomena. Such analytical reductionism characterizes science today. It led to amazing successes in the Industrial, Chemical, Electrical, Electronic, and other scientific revolutions. The necessity, however, for coupling analysis with synthesis is only just beginning to be recognized today.

SUGGESTIONS FOR FURTHER READING

Aristotle. *Of Generation & Corruption.* London: Oxford University Press, 1922.

Bernal, J. D. *Science in History.* Cambridge, Mass.: MIT Press, 1971, ch. 4: "The Iron Age: Classical Culture."

Clagett, Marshall. *Greek Science in Antiquity.* New York: Collier Books, 1963.

Farrington, Benjamin. *Greek Science.* London: Pelican, 1953, part 1.

Gershenson, D. E., and D. A. Greenberg. *Anaxagoras and the Birth of Scientific Method.* New York: Blaisdell, 1964.

Graubard, Mark. *Astrology and Alchemy.* New York: Philosophical Library, 1953, book 2: "Alchemy."

Heath, Thomas. *Aristarchus of Samos, The Ancient Copernicus.* New York: Dover, 1981.

Hogben, Lancelot. *Sciences for the Citizen.* London: Norton, 1951, part 2: "The Conquest of Substitutes."

Lucretius. *The Nature of the Universe.* Baltimore: Penguin, 1964, book 1: "Matter and Space," book 2: "Movements & Shapes of Atoms."

Plato. *Timaeus,* translated by Francis M. Cornford. New York: Bobbs-Merrill, 1959.

Sarton, George. *A History of Science.* New York: Wiley, 1964, chs. 4, 7, 8, 10.

Seneca. *Physical Sciences.* New York: Macmillan, 1910.

Van Melsen, Andrew G. *From Atomos to Atom.* New York: Harper, 1960, part 1.

4

Early Observation, Reference Frames, and Relativity of Motion

Ah, but a man's reach should exceed his grasp,
Or what's a heaven for?
                    Robert Browning, "Andrea del Sarto," 1855

The mind of man, urged on
By an invisible passion never will cease
To ask, "What is Beyond?"
                    Edward Fitzgerald, *Rubaiyat of Omar Khayyam,* 1859

MOTION EVERYWHERE

We live in a dynamic universe. Everything is ceaselessly moving or changing. Electrons, the smallest units of matter, are said to spin as they revolve around protons within atoms. Atoms either dart about freely as gases or vibrate within molecules. They combine and recombine continuously to form hundreds of thousands of different molecules. Molecules themselves seemingly are always in random motion, whether in solids, liquids, or gases. Or they in turn break down into atoms and reform into hundreds of thousands of changing substances.

This constant motion is not always apparent. The book in your hand may seem to be at rest. But the impression is only illusory. The molecules within its pages are vibrating; some are uniting with oxygen in the slow aging of the paper. Further, the book itself is moving at a fantastic rate of speed: It is carried by the earth spinning daily on its axis and orbiting at 67,000 miles an hour around the sun. The fact that you and everything else on the earth travel together at the same rate gives the illusion of rest.

Winds and swift waters relentlessly wear down the land; earthquakes and volcanoes force it again above the sea. Individual plants, animals, and human

59

beings live and die in groups that undergo evolutionary change. In the heavens the planets, moons, suns, and stars revolve endlessly in orbits of their own—much on the pattern of the tiny electron. Nothing ever seems free of motion.

In so dynamic, vast, and seemingly capricious a universe, man has always needed a measure of order, prediction, and accommodation. Without these, he is reduced to insignificance. Worse than that, without some basis for anticipating and contending with adverse natural events, man could not long survive in a universe of motion and change.

THE RHYTHM IN THE SKY

Man's earliest search for order was in the heavens. Little wonder! The heavens are a vast natural laboratory open to any creature that can observe, wonder, think, and record. The beauty, sparkle, and mystery of this immense laboratory are now being explored by rockets and spaceships. But even the earliest of men saw that motion in the heavens was not chaotic. Motion and change there were generally periodic and regular. It was perhaps in the observation of regularity in the motion of celestial objects that astronomy as a science was born.

In addition to the sun and moon, men observed myriads of stars daily rising and setting in fixed patterns. Each star retraced its own particular path in the sky. The ancients imaginatively translated these patterns into easily comprehensible terms and gave them life in the forms we know as constellations—Taurus the Bull, Leo the Lion, Scorpius the Scorpion, and many others. Ancient men also observed that these stars daily revolve around a central star, or pole star. All such visualizations led to conceptual model building on an ever larger scale.

Continued observation revealed five "stars" in addition to the sun and moon, apparently wandering in and out of fixed constellations. The Greeks applied their word *planets,* meaning "wanderers," to the bodies we know today as Mercury, Venus, Mars, Jupiter, and Saturn and to the sun and moon. Soon the first five were seen as decidedly different in motion, size, and brightness from the other two, and celestial objects were grouped into fixed stars, planets, sun, and moon.

As early as 3000 B.C.E. in Babylonia, movements of heavenly bodies were charted accurately enough for the astronomers of that period to develop calendars. (See chapter 2.) Regularity of celestial motions thus meant that time could be systematized and that events could be predicted on a natural basis. A calendar associated periodic floods with regular motions of sun and stars. Such predictability meant large-scale adjustment to surroundings and tremendous advantages for Egyptian agriculture.

There can be no science without observation and no observation without frames of reference. To demonstrate how this is so, consider specific patterns of celestial motion such as that of the daily westward motion of celestial objects, the eastward drifts of the sun and moon, and the retrograde (reversed) motion of planets.

Most obvious is the general east-to-west motion of celestial objects. The sun, moon, planets, and stars rise daily in the east, trace a semicircle across the sky, and disappear beneath the western horizon. The sun clearly brings the day; the moon often reduces the darkness of the night. There are no exceptions to this uniformity of direction. All objects in their daily half circles seem to move from east to west, none from west to east with respect to the horizon. And what is more, the sun, the moon, and all the planets apparently move across the same narrow belt of the star-studded sky.

In a sense, the heavens are like an open umbrella with white dots. It is as if the umbrella shaft is stretched horizontally across a table, with the white-spotted top hanging over the end of the table, half in and half out of sight. When the umbrella is rotated, the tabletop becomes the observer's horizon, across which the patterns of dots rise and set, unfailingly and always in the same direction.

As an observer travels north on the earth's surface, the North Star appears ever higher above the horizon in accordance with the latitude of the observer. In San Francisco, for example, at 37° north latitude, the North Star appears at 37° above the northern horizon. Or from Chicago, 42° north latitude, the North Star appears at 42° above the northern horizon. With this increase in altitude of the North Star above the horizon, more and more stars around it that formerly disappeared under the horizon can be observed making complete circles around this central star. For an observer at the North Pole, no stars would rise and set. He would see all the stars moving in complete circles around the polar star like dots in a mammoth umbrella rotating directly overhead.

The location of an observer thus partly determines what he sees. Everyone looking at the heavens has a frame of reference that must be taken into account in scientific observation. In the above cases, the *observer's frame of reference* is his position on the earth's surface, his latitude. This determines what he sees of the stars' apparent motions around the North Star or around his own horizon.

Of importance equal to the frame of reference of the observer is another related to it. This second frame of reference is a *background frame for the object observed.* In the daily motions of sun, moon, or stars described above, the North Star is a background point of reference for the sun, moon, or stars regardless of where the observer is in the Northern Hemisphere.

Now consider the yearly eastward drift of the sun in respect to the stars. Each morning the twinkle of any given star rising with the sun is quickly lost in the bath of the sun's light. Careful observation at dawn the next morning shows that now the same star rises about four minutes before the sun or that, conversely, the sun falls behind the star by about four minutes. If the sun makes a complete cycle in 24 hours, the star does so in 23 hours and 56 minutes. You can easily observe this phenomenon yourself. Check the position of any star at say, eight P.M. tonight. Check again tomorrow night, and you will observe the star to be at the same position at 7:56 P.M., since we base our 24-hour day and our clocks on the sun's apparent movement.

This daily falling behind of the sun with respect to the stars is known as the eastward drift of the sun. Thus, the sun appears to move *westward* with respect to the earth's horizon; but it does so more slowly than the stars by four minutes in every 24 hours and therefore appears to move *eastward* with respect to the stars. An analogy is two cars moving westward on a highway at 50 and 60 miles per hour, respectively. With respect to a ground observer, the slower car (the sun) is moving westward; with respect to an observer in the faster car (a star), the slower car is receding eastward.

On each succeeding day, the gap between a given star and the sun widens by an additional four minutes of time or 1° of space. The sun is about 2° behind the star in two days, 3° behind in three days, and so on. After 365¼ days, the sun will have fallen back 360° to the position of the given star again. The sun will have fallen one cycle (or lap) behind the star in its "race." The daily cycles of the star around the earth will have been counted at 366¼, while that of the sun will have been counted at 365¼.

You will recall that this interval between the two heliacal risings of the star, that is, between the two occasions when the sun and any given star rise together, was first arbitrarily defined as one year by the Egyptians. Thus, the eastward drift involves a yearly cycle. To summarize, there is the apparent original *daily westward* cycle of the sun and of the stars around the earth. This is in respect to the horizon as a reference frame. But because the sun slowly falls behind the stars in the daily cycles, the sun also has a *yearly* cycle of *eastward drift* with respect to the stars as all seem to move about the earth. Thus, whether "westward" or "eastward" depends upon the observed frame of reference chosen. The choice by the observer of the reference frame (horizon or stars) determines direction described (westward or eastward). We have relativity of motion then. And a relationship between the observer and object observed. And increased complexity in ancient conceptual building. (These ideas will be further developed in the "frames of reference" section below.)

The particular belt of stars through which the sun drifts in yearly cycles is called the zodiacal belt (see chapter 2). In this area are always found the same stars in patterns that were arbitrarily grouped by the Babylonians and then the Greeks into twelve constellations, or "signs" of the Zodiac. These constellations stretch along the belt at equal lengths of 30° for the total 360°.

"Zodiac" means "zone of animals"; all constellations in it except one, Libra (the Scales), appeared to the ancients to be figures of living creatures, such as the Lion, the Fishes, the Scorpion, the Bull, and the Water Bearer. The sun in its ecliptic slowly drifts eastward into and out of each one of these signs during one year. Thus, the zodiacal belt is really a background frame of reference of the sun in its eastward drift.

The moon, like the sun, daily appears to move *westward* with respect to the earth's horizon and *eastward* with respect to the moving zodiacal belt of stars. If the moon rises on any evening with a given star, on the following evening the moon will be 54.5 minutes behind the star instead of only four minutes late like the sun. The moon is found in about the same position in 24 hours and 50.5 minutes. The star is already 13.2° above the eastern horizon on the second evening when the moon shows itself. With the moon drifting behind the stars 54.5 minutes or 13.2° daily, it takes the moon 27.3 days—dividing 360° by 13.2° per day—to fall back to any given star. These 27.3 days constitute the cycle of eastern drift for the moon, as against the period of one year for the sun. That is, the moon rises (or sets) with any given star once in every 27.3 days, or about once a month.

Note again that the eastward drift pattern of the moon, like that of the sun, exists only in relation to a background frame of reference, the zodiacal belt of stars. Each of the planets also shows a general eastward drift in the zodiacal belt. Mars, for example, at the same hour from one evening to the next, falls behind any given star in the belt without fail—almost. It was this "almost" that made certain "wanderers" a special source of interest to ancient man, for from time to time, the eastward drift of five planets was reversed. During approximately two months in every 26, Mars was seen to reverse its direction from east to west relative to the stars. Instead of drifting further behind surrounding stars as expected, Mars strangely seemed to gain on them and even catch up. If plotted, this seemingly reversed motion relative to the stars forms a loop, as indicated in Fig. 4.1. Then, after about two months, Mars again settled back to the customary eastward drift for about two years. Such reversed motion, seen for planets only, is known as *retrograde motion*. The sun and moon do not show such motion in their zodiacal "wanderings," and so by the seventeenth century were removed from the (ancient) list of planets.

Note here also that planetary patterns of retrograde motion are observable only in relation to the reference frame of the zodiacal belt of stars.

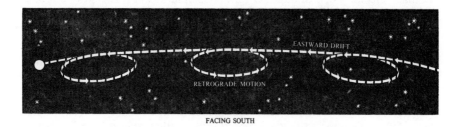

FACING SOUTH

Figure 4.1

Mars in apparent retrograde (reversed) motion.

FRAMES OF REFERENCE AND OBSERVER-OBJECT INTERRELATIONSHIP

The importance of frames of reference in astronomy warrants further attention. We repeat: there can be no science without observation, and no observation without frames of reference. In the previous section we observed motions of different celestial bodies in some detail. In so doing, we found it necessary to relate the moving bodies to other objects. The sun "rises" with respect to the eastern horizon and moves *westward* with respect to it; but the sun also has an *eastward* drift with respect to zodiacal stars as it falls behind them. Planets sometimes reverse their motions with respect to the stars but always move from east to west with respect to the horizon. *Motion cannot be observed except in reference to other objects.* Whether the sun moves eastward or westward depends upon whether its apparent motion is with respect to the stars, the horizon, or the moon. But if the direction and speed of an object depend upon the moving object to which it is referred—and all objects move—then all motion is relative. That is, in our dynamic universe of relative motion, any object moves with respect to a second object, which moves relative to a third, and so on. The moon, let us say, moves relative to the stars, which seem to move relative to the earth. Such a chain of relative motion could be broken only if the earth were stationary and the motion of all objects could be compared to it. How fast or in what direction any object is actually moving can be determined only if there is something in the universe that is at rest. The motion of other objects compared to it would then give a true, or absolute, motion. Thus, we must emphasize with Copernicus, in his *Revolutions of the Heavenly Bodies,* that "it is the Earth from which the rotation of the Heaven is seen." The earth is the astronomer's frame of reference, and what the astronomer sees depends partly on the earth's position and motion in space as well as upon his own position on the earth. In art or photography when the observer changes his position, he changes the background of the objects he observes. If, of course, the background should also be moving independently, what is photographed or observed is further changed. In astronomy it is the same: Each position on the earth has its own horizon; every motion or apparent lack of it is expressed in terms of the observer's location. The sun rises at one person's horizon after it has risen for a person to the east or before it does so for an observer to the west. This, of course, has resulted in twenty-four hourly time belts around the earth. Or you will recall that at different latitudes on the earth's surface, the North Star is seen at different altitudes above the horizon. This again is in accordance with horizons for different observers on a curved earth. Another example is that from the earth, the monthly path of the moon in its eastward drift against the zodiacal belt is a circle. If witnessed from the sun, the path of the moon around a revolving earth would appear more like a cycloid than a circle. Which observer would be correct—the one on the earth or on the sun? In an absolute sense, neither would be correct; in a relative sense, each would be correct from his own frame of reference. Observation becomes a relative relationship between the observer and the thing observed. It is like the case of a passenger in a moving automobile who drops a stone from the window. He sees the stone fall in a straight line. But a pedestrian at the side of the highway sees the stone fall in a parabolic curve. Whether the path of fall is a straight line or a parabola depends upon the position of the observer. Relativity of motion again! Similarly, the astronomer must be concerned with his own frame of reference, the earth and how its motion (or lack of it) and his position on it determine what he sees.

Actually even more important than the shape of the earth in determining what is seen from it is the question of whether or not the earth is moving. If the earth is stationary, the observations of the speeds, directions, and paths of all other objects are true ones. If the earth is moving, then "diurnal rotation is only apparent in the Heavens and real on Earth," says Copernicus in his *Revolutions.* We must therefore have something else fixed in the universe to determine our own position and velocity. We otherwise cannot make the proper allowances in determining the motions of all other objects and must accept all motion as relative. In other words, with all else in motion, only a stationary earth could provide the fixed frame of reference necessary for us to determine the true, or absolute, motions of all objects in the universe. It certainly would be of advantage to have the earth fixed, with all other objects in absolute motion around it.

Upon what basis, even against our immediate senses, was a stationary earth disclaimed and with it an earth-centered universe? Does the sun revolve around the earth or the earth around the sun? These are questions that led to the famous Ptolemaic-Copernican issue in which one conceptual model eventually was eclipsed by another. To answer them, observed specific motions in the heavens already discussed were crucial. They represented observations from the earth that any theory of the universe would have to take into account. In the process some insight into the nature, content, and structure of the universe as well as of science itself was gained.

CONCLUSION

Motion and change are characteristic of all things, but observation shows that, at least in the heavens, motions are not random or chaotic. And at least *from the standpoint of men on earth,* it has long been apparent that a regularity, a periodicity, exists in the motions of heavenly bodies. Things move in patterns or cycles. This regularity provides a basis for some kind of order; and order enables prediction, both prerequisites of science. Thus, astronomy became one of the earliest of sciences.

But we have also seen that there is no meaningful observation without two interrelated frames of reference: a background frame for the object observed and another frame for the observer. When characteristic motions of various heavenly bodies are considered in detail, a relativity of motion becomes obvious. That is, the specific orbit of any given object can be observed and described only relative to a background of other moving objects. The earth, our particular frame of reference as observers, also had to be considered with respect to its shape, and with respect to the question of its own motion. Whether the earth is fixed or moving is a particularly important question, for the earth influences what men observe and interpret in an apparent universe of motion.

SUGGESTIONS FOR FURTHER READING

Bernard, Hubert J., et al. *New Handbook of the Heavens.* New York: Mentor paperback, 1959, chs. 1–3.
Childe, V. Gordon. *Man Makes Himself.* New York: Mentor paperback, 1963.
Clagett, Marshall. *Greek Science in Antiquity.* New York: Collier paperback, 1963, ch. 1.
Farrington, Benjamin. *Greek Science.* Baltimore, Md.: Penguin paperback, 1961, ch. 1.
Hogben, Lancelot. *Science for the Citizen.* London: Unwin, 1951, chs. 1–2.
Kuhn, Thomas. *The Copernican Revolution.* New York: Vintage paperback, 1959, chs. 1–2.
Perlman, James S. *The Atom and the Universe.* Belmont, Calif.: Wadsworth Publishing Co., 1970, ch. 1.
Sarton, George. *A History of Science: Ancient Science through the Golden Age of Greece.* New York: Wiley paperback, 1964, chs. 1–3.

Part Two

Science as Conceptual Model Building, Projection, and Testing

5

Designing the Heavens: Geometry Shapes Astronomy

... [L]ike a floating iceberg whose bulk is largely hidden in the sea, only the smallest part of the physical world impresses itself upon us directly. To help us grasp the whole picture is the supreme function of [conceptual models].
Gerald J. Holton, *Foundations of Modern Physical Science,* 1953

From the earliest times men have used apparent groupings of stars as frames of reference to locate particular stars. You will remember, for example, that the Little Dipper was a frame of reference for the North Star. Or that the zodiacal belt was a projected frame of reference for motions of the sun, moon, and planets. As men came to accept the concept of a spherical earth, first claimed by the ancient Greeks, they established what is known as a system of coordinates to locate positions on models of the earth's surface. They halved the earth with the imaginary line that is the equator. They subdivided it further with lines of latitude as measurements of distance north and south of the equator.

They also sectioned the earth with longitudinal lines running north and south and subdivided it into degrees of longitude east or west of an arbitrary prime meridian. The 0° (or "prime") meridian we use today is the line running from pole to pole through the Greenwich Observatory, outside London.

Greek astronomers also projected upon the heavens a celestial sphere that completely encompassed the earth. After all, do not the heavens appear to be a huge dome? With the sphere was projected a celestial equator above that of the earth and a celestial prime meridian as coordinate axes for mapping the sky. As part of this coordinate system, a number of other reference points and lines were projected upon the celestial sphere directly above their earthly counterparts as celestial equator, poles, and lines of declination (latitude), or zenith, nadir, and celestial meridians. Such systematic mapping of the heavens was invaluable in the location of stars and in navigation.

According to Aristotle, "men are political animals." But Aristotle did not go far enough: Men are system-building animals; "political" organization is only one form of system building. In fact, as system builders, men are unique. Bees construct hives, and beavers dam streams. Both operate by instinct and senses to build systems of things. They do not learn these techniques, nor have they been capable of altering or adapting them over countless centuries. The habits of bee and beaver, ant and eagle, nut-gathering squirrel and bone-hiding dog are the same now as they were ten thousand years ago. Changes in habit occur only as the species themselves change.

But men have minds and imagination. They not only learn, adapt, and invent but also build systems of ideas as well as of things. Some of men's idea systems we know as philosophy, religion, literature, music, art, politics, and social theory. And, of course, we have houses, furniture, bridges, spaceships, and all the other wonders of technology resulting from men's building system of things. None of these tangible products, however, is possible without the ideas behind them. Science is a synthesis of a world of ideas and a world of things, and man is the synthesizer. Specifically, in astronomy, when men projected frames of reference and celestial spheres upon the heavens to locate moving objects there, they were scientifically building a system. They were projecting ideas upon nature in order to systematize nature. More than that, they were interacting with nature: Light signals from the stars moved inward; the celestial sphere was projected outward. The sphere was a joint project of interaction between man and his celestial surroundings.

Observation in science is basic but insufficient. Necessary also are ideas that relate and explain what is seen. With ideas we can relate what we observe; we can draw conclusions from our observations; we can explain and even predict events.

Gerald Holton, in *Foundations of Physical Science,* neatly expressed the role of theory in science with this remark: "The task of science as that of all thought is to penetrate beyond the immediate and visible, to formulate connections and thereby to place the observable phenomena into a new and larger context." But in astronomy, how can we attain such a larger context? Even with space travel, no man will be able to get above or beyond the universe to observe it from some neutral point.

We can start with ourselves and our earth. Just as the ancient Greeks did, we can project a celestial sphere to assist us in observing the stars. In that sphere we can establish a zodiacal belt in which we, too, locate the wandering planets, sun, and moon. These early models at least permitted careful observation and accurate measurement of time and place: where the heavenly bodies were, where they had been, where they will be and when.

Charting the movements of these objects, ancient people determined orbits and periodicity; upon these movements they calculated time, organized calendars, and made predictions. But working within frames of reference by observation, they learned only *how* objects move—not *why*. Like us today when we inquire *why*, they advanced not so much beyond as into the problem of *how*. To take such a step required forming a larger picture or a model of the universe that related observations in a unified, coherent manner. But how, then and now, do we obtain this more comprehensive model?

GEOMETRY SHAPES ASTRONOMY

The ancient Greeks let geometry provide the model much as we use other mathematical systems today. While they were studying the skies, the Greeks were also developing geometry into a very substantial mathematical system. Euclid brought this science to a pinnacle about 300 B.C.E. The Greek mind, with its interest in logic and the abstract, found rationality, order, and consistency in geometry. In fact, here were all the most desirable elements for a logical, understandable universe, predictable to human beings.

Pythagoras (fifth century B.C.E.) and his followers had been eager to fit a number system to the universe and then to proceed mystically, almost religiously, to ascribe a geometric order, truth, and beauty to the essence and structure of the universe itself. The numbers and geometric patterns became the underlying realities. After all, aren't spheres and circles characteristic of celestial objects and motion? Aren't spheres and circles perfect geometric figures? And wasn't perfection to be associated with the heavens? It was symbolically all there in the Pythagorean "music of the spheres."

In this way, it often became a simple matter to treat the celestial sphere less as a projected conceptual frame of reference of many moving points of light and more as a large, geometric, mechanically rotating structure of which the stars were a part. In other words, the stars would not be objects that moved and rotated on their own. Rather, the sphere became the unit, and the stars became points of light carried within it. Once a celestial sphere has been imagined and projected for observing the heavens, it is but one more step in theoretical system building to consider the sphere a mathematical reality. Besides, ideas were realities to the Greeks, as Plato expresses it so well with his cave analogy in the *Republic*.

This conception of a rotating sphere might again be compared to the large, transparent plastic umbrella previously described, on which the stars were only spots. But this time—more literally—there can be an actual sphere to do the rotating. According to Aristotle (384–322 B.C.E.), the "umbrella" can be composed of an ethereal crystal sphere made up of a fifth element, distinctly different from the terrestrial elements of air, earth, water, and fire. Or, according to a contrasting theory introduced about 373 B.C.E. by Hicetas, Ecphantus, and Heracleides, the celestial sphere and points of light can remain stationary while the spherical earth at its center keeps rotating. Because of relative motion, what would be observed from earth would be the same under either theory. To illustrate, have you ever been in a train watching another on a neighboring track when suddenly you observed it to be moving backward? Then, you look around and observe instead that your own train has gently begun to move forward.

An important consideration in all the above, however, is the fact that these ancient scientists had made an interpretation beyond observation in their attempt to answer the *why*. The celestial sphere originating as a convenient background frame of reference for locating stars became an imaginary theoretical structure in the heavens, a rotating geometric system. The rotating structure could be used to explain why, where, and when moving objects were seen. Such a theoretical system is an idea; it is a conceptual model or conceptual image of the world. Such a model, as theory, should be distinguished from the observations that suggest it and that are related to it.

From Plato's *Phaedo* comes this quotation: "This was the method I adopted: I first assured some principle which I judged to be the strongest and then I affirmed as true whatever seemed to agree with this, whether relating to the cause or to anything else; and that which disagreed, I regarded as untrue." (Doesn't this sound like a great deal of our personal, political, and sociocultural thinking today?) To further illustrate, let us quote Alfred Whitehead from his *Science and the Modern World*: "The Greeks thought that metaphysics was easier than physics and tended to deduce scientific principles from *a priori* conceptions of the nature of things. They were restrained in this disastrous tendency by their vivid naturalism, their delight in first-hand perception. [Later] Medieval Europe shared the tendency without the restraint."

With respect to astronomy, Plato advised further: Determine "what uniform and ordered movements must be assumed for each planet to explain its apparently irregular paths." This was a specific directive to his students in their attempts to call the heavens to order. Implied in his words is a sophisticated recognition that things as they appear ("apparently irregular movements of the planets") may not be as they actually are ("uniform and ordered movements"). Perfection was to be expected in the heavens. Yet retrograde (reversed) motions involved irregular, changing paths for planets that could be easily observed. The problem was this: How can the idea of perfection of the heavens be reconciled with the imperfection of what is seen? How can only circular motions or combinations of them be used to explain apparent motions of celestial objects that are not circular? To Plato, perfection of all things in the heavens was the reality to be reconciled with the imperfect shadows that we observe. The perfect form of geometry could do the job of reconciliation, but how, specifically? Plato meanwhile inscribed these words above the entrance of his Academy: "Let no one enter here unless he has a taste for geometry and mathematics." Mathematical harmony was the basic guiding principle to the universe. Mathematics is still considered a key to the universe today.

Of Plato's various students, Eudoxus (409–350 B.C.E.) developed perhaps the most ingenious answer to his master's problem of reconciling or "saving the phenomena." In considerable detail, he developed a model of rotating concentric spheres. The model explained various celestial motions by embedding sun, moon, and planets within separate rotating spheres enclosed by an outermost sphere of stars (Fig. 5.1). The fixed earth was at the center of these eight spheres, arranged like layers of an onion. All the spheres were assumed to be rotating at different speeds and in different directions around different axes. To account for retrograde motion, each planet needed several special auxiliary spheres in connection with its own to show how many interconnected objects in uniform circular motion can give the appearance of changing speed and direction.

Eudoxus ascribed a total of 26 simultaneous uniform motions—and Aristotle, later, 55—to account for the particular motions of the sun, moon, and planets. As discrepancies accumulated between the positions of planets predicted by the model and the actual observed positions, more auxiliary spheres added to the system eliminated the differences between prediction and observation. There was, however, an Achilles' heel in Eudoxus's world system. His model could not explain why planets appear to vary considerably in brightness as time goes by. No matter how many auxiliary spheres were used to account for apparent positions or directions of a planet, the planet itself would always in this system of concentric spheres have to be at about the same distance from the observer and, therefore, should always show about the same brightness. And it didn't. After all, the observer is at the center of the original sphere carrying the planet itself and, therefore, always close to an equal distance from it.

In spite of its weakness, Eudoxus's model of the universe was an important scientific development for the following reasons:

1. It was an ingenious conceptual device. Although his model was inadequate to attain his objectives of highly accurate predictions with the assumption of the perfection of the heavens, Eudoxus established a significant precedent of using mental models to explain, unify, and predict phenomena of nature.

2. By establishing such use of geometric models as bridges between astronomical theory and observation, Eudoxus stimulated such successors as Aristotle (also a student of Plato), Hipparchus, and Ptolemy to work for modifications or for more adequate models.

3. His model gave further promise for mathematics as a language of nature and of science.

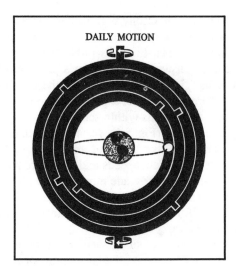

Figure 5.1

Concentric spheres model of the universe. The model explained various apparent celestial motions by embedding sun, moon, and planets within separate rotating spheres enclosed by an outermost sphere of fixed stars. Spheres were connected by axes rotating at various angles and velocities.

4. Through the model, explanations of events were possible in terms of natural causes. That is, most known astronomical events could be explained in terms of the immediate workings of nature itself.

5. Concretely and ingeniously, Eudoxus, through this model, spelled out differences between apparent and actual motions. Apparent motions were what was observed, including such irregularities as retrograde motion, whereas actual motions were represented by perfect geometric models, befitting the perfection of the heavens. Retrograde motion as an observation of imperfect motion was a combination of the actually perfect motion of the spheres.

AN EARTH-CENTERED MODEL

Greek astronomy reached its culmination in Claudius Ptolemy (100–178 C.E.). This meticulous scientist, like Hipparchus three hundred years earlier, set a pattern for all future astronomers in the extent and care of his observations. He cataloged 1,080 stars, almost half of those visible to the naked eye. More than that, using the earlier concepts of Hipparchus, Ptolemy developed in great detail from his observations a world system powerful in its ability to predict future motions of celestial objects with precision. His great synthesizing work, the *Almagest,* dominated astronomy until the time of Copernicus 1,400 years later. Ptolemaic tables are still used in navigation today.

The Ptolemaic world system (Fig. 5.2) is based on common sense.[1] Like Aristotelian cosmologists before him, Ptolemy placed the earth at the center of the universe, and although he acknowledged its sphericity, his earth was stationary. This is easy to understand. The earth appeared to be at the center of the universe and fixed; therefore, Ptolemy assumed that it was so. Similarly, the stars appeared to be revolving; thus, he assumed that they indeed were. Even today in common conversation we attribute movement to the sun when we say that it rises—and not that we are turning toward the sun. That appearances are deceiving is something we are still learning today. The problem is not an easy one: observation is still often confused with fact. Apparent motions of stars were thus interpreted by Ptolemy and other geocentric astronomers as *actual* motions.

Between the earth and the celestial sphere in the Ptolemaic model are the moon, Mercury, Venus, the sun, Mars, Jupiter, and Saturn in the order given. The circles, called *deferents,* represent the orbit of drift, with respect to the stars, of the various bodies around the earth, the observer's frame of reference.

To solve the basic problem of the planets, i.e., their varying brightness, however, Ptolemy replaced the concentric spheres of the planets with a mathematical representation of their orbits by circles, also a perfect geometric form. The use of circles solved the problem in the following way: Imagine a planet P to be orbiting in a circle around the point O' (Fig. 5.3). If meanwhile the center O' moves in a larger circle around a center O, the smaller circle is called an *epicycle,* and the larger, a deferent. If the deferent center O is not on the earth, the point O becomes an *eccentric* center in relation to the earth. Because of the displacement of the center from E, an object moving on the deferent would appear to be changing its distance from the earth. The object would be following an eccentric path. Still another center Q, called an *equant,* is an imaginary point from which the epicycle center O' moves at a uniform angular rate. That is, from it a given angle AQB will change at a constant rate as point O' revolves. Or if O' moves from A to B in the same time that O' later moves from C to D, the two angles formed, AQB and CQD, are equal. (That is, $\angle 1 = \angle 2$). The angle AQB changes at a constant rate as O' moves around the circle. *Constant angular speed* exists with respect to equant center Q and circular motion with respect to deferent center O. The assumption of uniform circular motion was thus satisfied while allowing the planets to vary in brightness. We shall also see that through the use of deferents, epicycles, eccentrics, and equants as geometric devices, the Ptolemaic model, at the time, could predict motions of celestial bodies with adequate precision.

All ideas, beliefs, theories, or statements rest upon assumptions of some

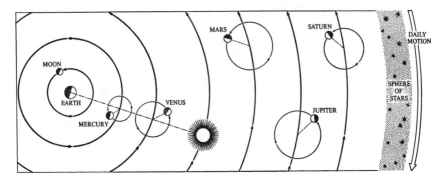

Figure 5.2

Plotemaic world system. Ptolemy replaced planetary concentric spheres with circles representing planetary orbits around the earth.

kind. Since no idea can be stronger than the assumptions it rests on, assumptions are extremely important in the examination of ideas. In fact, science in large part, as we shall see, is a matter of testing assumptions. We can consider the Ptolemaic model to have been based upon the following main assumptions. Relate each of these assumptions to the Ptolemaic diagram in Fig. 5.2.

1. A stationary, spherical earth exists at the center of the universe. There are actually three assumptions here, based on observation and Aristotelian cosmology: a fixed earth, a spherical earth, and the earth's location at the "center of the heavens."

2. All heavenly bodies move in perfect circles or in systems of perfect circles around the earth. This assumption was in tune with prevailing ideas associating the heavens with divinity and with mathematical perfection.

3. All circular motions of celestial bodies are at constant angular speeds. This assumption was also in line with the Aristotelian doctrine of the perfection of the heavens.

4. All "fixed" stars move in one outermost sphere. This assumption was also based on observation. The difficulty of adequately observing the third dimension, depth, permitted the placement of all stars in a single sphere. Observations of unchanging angles among stars as they apparently moved in groups also substantiated this assumption at the time.

5. The sun, moon, and planets occupy their own individual orbits daily around the earth. As with Eudoxus's assumption of concentric spheres, more careful

observation (for example, noting that one body passes in front of another in transits or eclipses) justified removing the sun, moon, and planets individually from the sphere of stars.

6. The individual orbits of sun, moon, and planets are at considerable distances from one another. By the time of Ptolemy, geometric technique had become far enough advanced to give a rough idea of this distance. In the third century B.C.E., for example, Aristarchus had already used triangulation methods with the sun, moon, and earth to determine relative distances of the first two from the earth.

7. The size of the earth compared to its distance to the sphere of stars is so insignificant as to permit treating the earth as a mathematical point in the universe. Here again, the application of geometry to astronomy gave ever increasing respect to distance in astronomy as compared to the size of objects.

In examining the above assumptions from the vantage point of today's knowledge, only the last assumption is without fallacy. The first six, one by one, were based upon observation, beliefs, or mathematical reasoning but have not lasted. Obviously, the senses, the mind, or the imagination alone is not enough for the advancement of knowledge and understanding of the universe. We need to use everything we have, senses, mind, and imagination, in a system of checks and balances for a working, solid grasp. It is a matter of using, through time, everything we have with more and more precision of observation, testing and refinement of ideas, and fertility of imagination. Mathematics in the physical sciences early became intimately involved with accuracy of observation or measurement, with systematizing and with reasoning.

We have previously discussed conceptual models as theoretical systems or schemes. A first main function of such models is to systematize and explain data already known. A second is accurately to predict future events. Such models also serve as conceptual reference frames like the earth-centered Ptolemaic model rather than reference frames of observation like the horizon. In performing these functions, conceptual models must be comparatively accurate, simple, reasonable, flexible, and fruitful of new developments. Otherwise, as we shall see, they do not survive. Let us therefore return to the basic astronomical observations already discussed. Our immediate purpose is to see how effectively the Ptolemaic model systematized, explained, and predicted these particular phenomena.

The Ptolemaic model easily explained the daily east-to-west motion of all celestial objects. The general diagram of the Ptolemaic system (Fig. 5.2) is from the vantage point of an observer on the north celestial pole. All objects move in the same daily *clockwise* direction around the earth. The celestial sphere carries all the stars imbedded in it. The sun, moon, and planets move

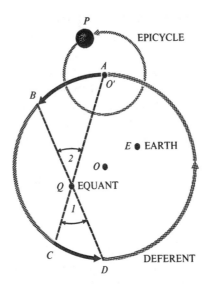

Figure 5.3

Epicycles, eccentrics, and equants reconciled observation and theory. These geometric devices involved three different centers around which planets moved.

as if carried in separate, inner spheres of their own. Their paths are also clockwise ones daily around the earth. (The counterclockwise arrows in Fig. 5.3 represent eastward drift, which is explained below.)

You will recall that the sun in its observed daily cycle around the earth falls behind any given star to the extent of about 1° of space or four minutes of time daily. If the sun rises with a given star today, tomorrow it will rise behind the star. How did the Ptolemaic model explain this eastward drift of the sun with respect to the stars?

In this model the sun has two cycles of motion. First is a daily cycle as its sphere carries it clockwise around the earth. Second is a cycle of eastward drift of the sun against the motion of its sphere as the sun moves about 1° a day within its sphere (Fig. 5.4), much like a ticket taker on a spherical merry-go-round. Creeping about 1° a day in this way, it takes the sun about 365¹/₄ days, or 365¹/₄ rotations of its sphere around the earth, to return to its original position in its sphere. The eastward drift of the sun in its ecliptic thus becomes a yearly cycle even though its sphere carries it daily around the earth. In all this the Ptolemaic model was successful. It explained motions observed; it afforded a mathematical basis for day-to-day predictions.

The Ptolemaic explanation of the observed eastward drift of the moon is similar to that of the sun's drift. The moon is also treated as if carried around clockwise daily in its own sphere. As we have explained, however, the moon, moves against the motion of its sphere to the extent of 13° a day and therefore loses 54.5 minutes a day. Every 27.3 days, the moon falls back to its original position within the sphere. This explains the faster 27.3-day cycle of the moon's eastward drift as against the 365$^1/_4$ days of the sun. This path of the moon's eastward drift traced within its sphere is also that with respect to the sphere of stars and constitutes the deferent of the moon. Here again, the Ptolemaic model successfully explained and predicted observed motions.

An explanation was also necessary for the retrograde motion of the planets. You will recall that this motion is a temporary reversal of eastward drift of the planets to a westward gaining on the stars. At this time the planets are observed to catch up to stars west of them rather than to fall behind.

For a planet actually to reverse itself in this way would be contradictory to the Ptolemaic assumption of perfection of the heavens. Celestial objects must consistently move in circles and at constant speeds. How were the irregularities of retrograde motion to be reconciled with these Ptolemaic assumptions? Once again, how were the "phenomena to be saved"? Mind you, it wasn't the guiding principle of perfection that was in jeopardy, but the phenomena of observation. Ptolemy's answer was Hipparchus's ingenious one of epicycles. The use of this geometric device would permit the planets actually to move in a system of perfect circles at constant speed and appear to retrogress. A point on the rim of a wheel rolling at constant speed always moves in a circle around the hub at constant speed. When this point approaches the ground for a short time, it can appear to be moving backward even though the wheel as a whole moves forward. This reversal can be photographed by attaching a lamp on a point of the rim of a wheel properly on a rotating disk.

For the Ptolemaic explanation of retrograde moon, place a planet on the rim of a small merry-go-round and call the planet's circular path an epicycle (Fig. 5.4). Then place the center of the small merry-go-round on the rim of a large moving merry-go-round and identify the circular path of the epicycle center as the cycle of eastward drift of the planet in the zodiacal belt. The planet, like any point or lamp on a wheel, will show a retrogression that is really a compound of two uniform circular motions. Uniform circular motion is defined as circular moon at constant speed.

Here Ptolemy and his followers, in their passion for perfect order, made the universe conform to that passion. Ptolemy based his thinking on Plato's guiding principle of perfection of the heavens, just as we today base our thinking on other guiding principles that may be rejected in the future. Ptolemy attributed the irregular motions of the planets to their supposed epicyclic movements and ingeniously reconciled these imperfect movements with prevalent perfect-circle theories.

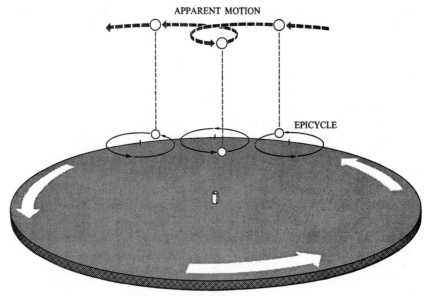

ACTUAL MOTION BY PTOLEMAIC THEORY

Figure 5.4

By Ptolemaic theory, retrograde motion is an apparent irregular motion of a planet that is really a compound of two or more uniform circular motions in an epicycle and deferent.

The Ptolemaic model of the universe was a dominant, successful theory for fourteen centuries. Why was it successful?

1. The Ptolemaic theory was based upon immediate sensory impressions or appearances—it contained so much "common sense." The earth certainly seemed to be fixed, and the heavenly bodies seemed to move around it.

2. The system exerted a strong psychological appeal to man's necessary egoism and unnecessary vanity by putting man and his earth at the center.

3. The system was aesthetically pleasing because of its coherence, unity, and symmetry.

4. A perfection-of-the-heavens doctrine exerted considerable philosophic-religious appeal, whether reflected in the Pythagorean Brotherhood, Plato, the Old Testament, or Milton's *Paradise Lost*.

5. Its scientific appeal was also strong. The Ptolemaic theory seemed to satisfy criteria of a good theory even today. It provided a working model for the universe that offered a neat, unified explanation and a basis for prediction of movements with built-in means of self-correction and modification. The theory had a definite sophistication in distinguishing between the observed motion and the (supposed) actual motion of celestial bodies. The difference is one between what *is seen* and what *is,* between *data* and *facts*—in today's terms, between arriving *signals* and actual *events* behind them.

By systematizing what was observed through the use of well-ordered celestial circles, the theory successfully met all challenges, even to account for the unusual phenomenon of varying brightness of the planets. In using epicycles, eccentrics, and equants as special flexible geometric devices for reconciling assumptions and observations, the Ptolemaic model was able to unify and to predict with considerable detail, precision, and self-correction. And with our hindsight, it led to the more adequate model of Copernicus. Thus, it satisfied prerequisites for any good scientific theory.

It is easy for us today in the late twentieth century to look back upon Ptolemy and his followers condescendingly. After all, we know the absolute truth! Or do we? Does his well-ordered geocentric universe look ridiculous to us? Well, how will our theories of the universe appear to scientists in the year 2500? Even today, in accordance with Einstein's theory of relativity, whether the earth revolves around the sun or the sun around the earth depends upon where we are as observers.

We respect Ptolemy for his great contributions to science and astronomy in terms of his period in history. He lived 1,800 years ago, over 1,400 years before the invention of even so early and relatively simple a scientific instrument as the Galilean telescope. Using his unaided sight only, Ptolemy was able meticulously to classify, to locate, and to describe accurately the apparent movement of some 1,080 or more celestial bodies. If he made interpretations that we do not accept today, we can understand why. From what point of reference was he able to observe the turning of the earth? How could he or any man, dependent upon senses alone, comprehend the turning earth? In the light of the limited knowledge of astronomy in Ptolemy's day, the apparent motions of the heavenly bodies were more explainable by a stationary earth. His geocentric model of the universe represents an earlier stage in the development of the same science that later efforts have brought to a more advanced stage. Men had to exhaust the strong commonsense possibilities of an earth-centered universe before seriously entertaining present ideas of an ever-expanding universe with the earth a mere moving speck in all this vastness. Thus we can see that science is a permanent, dynamic, and expansive process involving interaction between man and his surroundings that dates back to ancient ancestors and civilizations.

NOTE

1. While descriptions of Ptolemaic theory are based on Ptolemy's *Almagest,* figures in this chapter are taken from James S. Perlman, *The Atom and the Universe* (Belmont, Calif.: Wadsworth Publishing Co., 1970).

SUGGESTIONS FOR FURTHER READING

Bernal, J. D. *Science in History.* Cambridge, Mass.: MIT Press, 1971, ch. 4: "The Iron Age: Classical Culture."
De Coulanges, Fustal. *The Ancient City.* Garden City, N. Y.: Doubleday, 1956.
Dreyer, J. L. E. *History of Astronomy From Thales to Kepler.* New York: Dover, 1953, ch. 9.
Graubard, Mark. *Astrology and Alchemy.* New York: Philosophical Library, 1953, chs. 1–6.
Holton, Gerald. *Concepts and Theories in Physical Science.* Reading, Mass.: Addison Wesley, 1973, ch. 1: "Ancient Greek Astronomy."
Milton, John. *Paradise Lost,* book 8.
Pannekoek, A. *History of Astronomy.* New York: Interscience Publishers, chs. 9–15.
Sarton, George. *History of Science through the Golden Age of Greece.* New York: Wiley, 1964, chs. 16–20.

6

The Copernican Dilemma: Replacing a Model

The moving Finger writes;
 and having writ;
Moves on. . . .
                Edward Fitzgerald, *The Rubaiyat of Omar Khayyam*

To doubt this [Ptolemaic] system and to seek for another and better one at a
time when all men's minds were governed by tradition and authority, and
when to doubt was sin—this required a great mind and a high character.
                Sir Oliver Lodge, *Pioneers of Science*, 1893

A SUN-CENTERED MODEL

With the fall of Greco-Roman civilization, Western science was lost in the bar-
barism of the Dark Ages and the otherworldliness of Christianity. Astronomy
did not advance for about 1,400 years after Ptolemy. In 389 C.E., Theophilus,
Patriarch of Alexandria, ordered the pagan temple and library of Serapia
burned to the ground. This library had been well known in the classical world.
The even more famous library of Alexandria was completely destroyed by
Moslems in 641 C.E., but not before it had been badly depleted by Christian
drives against paganism.

Fortunately, copies of Ptolemy's *Almagest,* together with works of Aristotle and Plutarch, had found their way into the hands of appreciative Arab
scholars who translated them into their own language. Thomas Aquinas (1225–
1274) scholastically reconciled Christian theology with Greek science and philosophy. By this time, a number of churchmen had been retranslating Aristotle, Ptolemy, and other Greek authorities from Arabic into Latin. Then, with
the Renaissance, came a humanistic search for classics in the original Greek.

83

This brings us to Nicolaus Copernicus (1473–1553), born in the Polish town of Thorn, near the Prussian border, thirty years after the invention of the Western printing press.[1] While Columbus was busy with his own earth-shaking discovery of America, Copernicus was studying at the Universities of Cracow and Bologna. The young Copernicus took courses in law, medicine, astronomy, mathematics, and Greek. In 1503, his uncle, the bishop of Ermland, appointed his well-educated young nephew, who now also had the degree of Doctor of Ecclesiastical Law, as a canon in the Cathedral of Frauenburg. As a church dignitary and assistant to his uncle in governing Ermland, Copernicus was able to devote considerable time to astronomical observation and theory.

Knowing Greek and Latin, Copernicus could read the ancient texts directly. In establishing a sun-centered system of the world, he used Ptolemy's recorded observations as well as observations of his own. Copernicus worked quietly and published his great work, *De revolutionibus orbium coelestium* (*On the Revolutions of the Celestial Spheres*), at the very end of his life. The first copy of this book was brought to him from the printer the day he died.

Although we give credit to Nicolaus Copernicus for developing the concept of the sun-centered universe and a revolving, rotating earth, he was not the first to hold such theories. In Greece, four centuries before Christ, Heracleides of Pontus had introduced the idea of a rotating earth, as we have already noted. However, he placed his earth at the center of the universe with the planets Mercury and Venus revolving about the sun, which in turn revolved around the earth.

A century later, Aristarchus of Samos became an "ancient Copernicus" by positioning the sun at the center and sending his earth in an orbit about it, rotating as it revolved. But his ideas did not prevail against Aristotle, a contemporary of Heracleides, or against Ptolemy, four hundred years later, for several probable reasons:

1. From what we understand of the Aristarchan theory—the records are meager—Aristarchus handled his proposition only qualitatively. That is, unlike Copernicus much later, he did not base his theory on quantitative detail involving observations, tables, and predictions.

2. Aristarchus also did not have the benefit of a telescope to improve observation of the heavens. Technology had not yet developed to the stage where a bold new theory could be more readily supported (or rejected) by evidence from such an invention.

3. Hipparchus and Ptolemy, unlike Aristarchus, made elaborate use of mathematics and observation to establish their earth-centered system in a quantitative treatment. Their theory, grounded in tables, charts, and predictions, was able to provide a navigation aid to sailors. Aristarchus was therefore forgotten.

Our ideas of the original Copernican theory are based upon three of Copernicus's astronomical publications: *The Commentariolus* (a lengthy letter about celestial motion), a *Letter against Werner,* and *The Revolutions.* The first two are lengthy letters sent by Copernicus to friends who quietly copied and circulated them. They are really preludes to his monumental *Revolutions.*

Copernicus had great respect for Ptolemy and his work. He used Ptolemy's observational data without ever questioning its accuracy. And Copernicus recognized the great value of the Ptolemaic tables based upon his predecessor's data. Through these tables, men could still predict future positions of the moon, planets, and stars; and navigators like Columbus could sail far and wide on a spherical earth. But Copernicus also knew that the Ptolemaic system would be inadequate for predicting celestial motions in his own day unless over seventy circles were used, mostly epicycles upon epicycles upon epicycles, etc. Copernicus was too convinced aesthetically and religiously of a mathematical simplicity and harmony in the universe to be satisfied with so complicated a system, even if it was successful. He had read what there was about Aristarchus and other ancient writers; he believed that a sun-centered system would be more simple and appealing.

Copernicus also deeply respected the Ptolemaic assumption of perfection of the heavens, and so he believed that all moving astronomical objects definitely must orbit in perfect circles and at constant speeds. Concentric crystal spheres could well provide the celestial mechanism for this kind of movement. If the earth is moving, as proposed by Aristarchus, it is part of the universe and should be sphere-borne to conform to the master plan. It was not, therefore, the assumption of perfection that disturbed Copernicus. On the contrary, he thought that Ptolemy's equant violated perfection and harmony. As Copernicus himself says in the *Commentariolus*:

> Yet the planetary theories of Ptolemy and most other astronomers, although consistent with the numerical data, seemed likewise to present no small difficulty. For these theories were not adequate unless certain equants were also conceived; it then appeared that a planet moved with uniform velocity neither in its deferent nor about the center of its epicycle. Hence, a system of this sort seemed neither sufficiently absolute nor sufficiently pleasing to the mind.

But perhaps it would be best if, at this point, we let Copernicus describe his own system. The following famous excerpt is from the *Revolutions:*

> The first and highest of all the spheres is the sphere of the fixed stars. It encloses all other spheres and is itself self-contained; it is mobile; it is certainly the portion of the universe with reference to which the movement and

positions of all the other heavenly bodies must be considered. If some people are yet of the opinion that this sphere moves, we are of a contrary mind; and after deducing the motion of the earth, we shall show why we so conclude. Saturn, first of the planets, which accomplished its revolution in thirty years, is nearest to the first sphere. Jupiter, making its revolution in twelve years, is next. Then comes Mars, revolving once in two years. The fourth place in the series is occupied by the sphere which contains the earth and the sphere of the moon, and which performs an annual revolution. The fifth is that of Venus, revolving in seven months. Finally, the sixth place is occupied by Mercury, revolving in eighty days.

In the midst of all, the sun reposes, unmoving. Who, indeed, in this most beautiful temple would place the lightgiver in any other part than that whence it can illumine all other parts?

And now with respect to his "deducing the motion of the earth," Copernicus states in the *Commentariolus:*

Having become aware of these defects, I often considered whether there could perhaps be found a more reasonable arrangement of circles, from which every apparent inequality would be derived and in which everything would move uniformly about its proper center, as the rule of absolute notion requires. After I had addressed myself to this very difficult and almost insoluble problem, the suggestion at length came to me how it could be solved with *fewer and much simpler constructions* than were formerly used if some assumptions (which are called axioms) were granted me. They follow in this order.

1. There is no one center of all the celestial circles of spheres. [That is, the centers of the various spheres could be in various eccentric positions around the center of the universe.]

2. The center of the earth is not the center of the universe, but only of gravity and of the lunar sphere.

3. All the spheres revolve about the sun as their midpoint, and therefore the sun is the center of the universe.

4. ... The distance from the earth to the sun is imperceptible in comparison with the height of the firmament. [That is, imperceptible in comparison with the distance from the earth or the sun to the stars.]

5. Whatever motion appears in the firmament arises not from the earth's motion. The earth ... performs a complete rotation on its fixed poles in a daily motion, while the firmament and highest heaven abide unchanged.

6. What appears to us as motions of the sun arise not from its motion but from the motion of the earth and our sphere, with which we revolve about the sun like any other planet. The earth, then, has more than one motion.

7. The apparent retrograde [motion] and direct motion of the planets arise not from their motion but from the earth's. The motion of the earth alone therefore suffices to explain so many apparent inequalities in the heavens. (Original italics)

And then in his preface to the *Revolutions:*

... I found after much and long observation that if the motions of the other planets were added to the motions of the earth [its daily rotation on an axis and yearly revolution around the sun] ... not only did the apparent behavior of the others follow this, but the [new] system so connects the orders and sizes of the planets and their orbits, and of the heaven, that no single feature can be altered without confusion among the other parts and in all the universe. For this reason, therefore ... have I followed this system.

The seven assumptions listed above differentiate Copernicus's heliocentric system from Ptolemy's earth-centered system. Two further very important assumptions of Copernicus, however, must be mentioned; they are the same as for Aristotelian-Ptolemaic theory. These are the assumptions that all actual motions in the heavens must involve (1) perfect circles and (2) constant speeds.

With the first seven assumptions, Copernicus paves the way for a scientific revolution; with the last two, he remains tied to the past. He was a man in transition from the Middles Ages to modern times.

In summary, remove the equants from the Ptolemaic system, and the Copernican system becomes much like the Ptolemaic with the positions of the sun and earth interchanged. The moon, of course, accompanies the earth as a true satellite. Celestial spheres remain, as do epicycles and eccentrics. Deferents are still there but have become "yearly orbits" around the sun. True, the earth has been removed from the center of the universe, but not far removed; for the distance from the earth to the sun is almost negligible compared to that between the earth and the sphere of the stars. Besides, the sun, to Copernicus, belongs at the center. As the source of all light, the sun should be placed there for greater simplicity and harmony in the heavens. Moreover, with the sun in the center of the universe and the earth circling around it, the despised equant could be eliminated.

A crucial test of any theory is its ability to explain and predict specific events. Let us now see how the original Copernican theory could account for the basic observations of the heavens already described.

"It is Earth from which the rotation of the Heavens is seen," said Copernicus in his *Revolutions*. But the earth is the observer's frame of reference. This frame of reference actually is rotating on its axis every 23 hours and 56 minutes counterclockwise, from west to east, and of course, all observers travel with it. But by relative motion, the sun, moon, stars, and planets appear to move in an opposite direction around the earth from east to west and around the North Star counterclockwise. The North Star becomes the stars' point of reference because this star is almost in line with the axis of the earth and the zenith of the earth's North Pole. By way of illustration, a child rotating counterclockwise under the dome of a huge, stationary umbrella will see dots on the umbrella seemingly move clockwise with respect to himself but counterclockwise around the central, uppermost point of the umbrella. Thus, says Copernicus in the *Revolutions,* "the diurnal rotation is only apparent in the Heavens, but real [actual] on earth." And relative motion makes it difficult for an observer to distinguish between apparent and real motion.

It is obvious that both Ptolemy and Copernicus possessed adequate mechanisms of explanation and prediction for daily rotation of the heavens. Ptolemy had all objects carried in circles around the earth; Copernicus had the earth rotating instead. The only advantage that Copernicus could claim here is one of simplicity: Just rotating the earth dispenses with the need for the sun, moon, planets, and stars to make complete daily circles in the heavens. To the Ptolemaic concern that a rotating earth would fly apart because of its speed, Copernicus responded, "What about the celestial sphere of stars?" Under the Ptolemaic theory, the celestial sphere would be required to revolve through much greater distances at fantastic speeds to make a complete revolution in 24 hours. As for the anticipated Ptolemaic argument that with a rotating earth, relative winds would blow everything off the face of the earth, Copernicus's answer was that there is no problem: There is no relative wind because the atmosphere turns with the earth. Why this is so Copernicus would not have been able to answer. Galileo and Newton had not yet arrived with a new concept of gravity and a new science of mechanics. Followers of the Ptolemaic theory, however, did have Aristotle's mechanics to support an earth-centered universe, as we shall see later.

The eastward drift of the sun in its ecliptic was easily predicted and explained by Copernicus. The earth is not only rotating on its axis but also revolving around the sun at the rate of about 1° a day, or 360° a year (Fig. 6.1). Both the sun and the stellar sphere are stationary. This revolution of the earth results in the sun's being seen against a slightly changed background of stars after each rotation. Since the earth's revolution is counterclockwise, or in the same direction as its rotation, it has to continue rotating 1° for four minutes of time more than a complete circle for the next sunrise. This extra amount of turn accounts for eastward drift.

If the earth revolves around the sun, a closer star should appear to move against the background of further stars. This apparent motion is called *stellar parallax.* Astronomers from Aristarchus to Copernicus, Galileo, and others

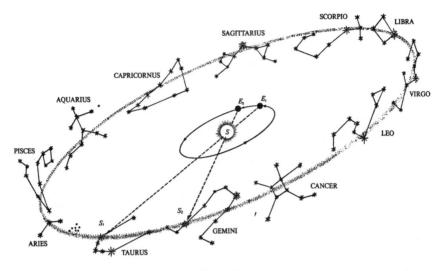

Figure 6.1

Parallax of the sun. The Copernican explanation of the sun's apparent eastward drift among the stars involves solar parallax. As the earth moves from E_1 to E_2, the sun appears to move eastward from star S_1 to S_2 in the background.

much later, looked for stellar parallax, but it was not to be found. To Ptolemaic astronomers, this was evidence that the earth does not revolve. The answer of Copernicus, as well as of Aristarchus long before him, was that stars are too far away for detection of annual parallax. The naked eye cannot detect it if it is there.

It wasn't until 1838 that stellar parallax finally was discovered officially by the astronomer Friedrich Wilhelm Bessel. The telescope had become a powerful enough instrument for Bessel to determine slight apparent shifts of the star 61 Cygni. Earlier astronomers had been handicapped by astrolabes or by telescopes of less resolving power and therefore were unsuccessful in discovering parallax. The importance of the development of tools and apparatus in the history of ideas in science cannot be overemphasized.

In the sixteenth century, however, before the discovery of parallax, Copernican and Ptolemaic theory could about equally well explain and predict the eastward drift of the sun. Certainly in this regard, the Copernican approach had no decisive advantage. And even if parallax had been observed at that time, Ptolemaic astronomers could have attributed it to epicyclic motions of the particular stars in the stellar sphere. And with respect to simplicity, Copernicus's earth, carried by a sphere yearly around the sun as it also rotates, is hardly simpler than Ptolemy's sun, slowly moving within its sphere as its sphere carries it daily around the earth.

The Copernican theory explains the eastward drift of the moon mainly through three motions occurring at the same time:

a. the earth's counterclockwise rotation on its axis,
b. the earth's counterclockwise revolution about the sun, and
c. the moon's counterclockwise revolution around the earth.

Looking at Fig. 6.2, let us start on a given evening, say May 10, with the earth at E_1 and the moon at M_1. If the observer is at O, he will see the moon rising. As the earth rotates once daily through 360° with respect to the stars, the moon meanwhile moves about 13° or 54.5 minutes of time for the same observer to sight again the moon rising—but it is now against a different background of stars. The moon seems to have fallen behind or drifted to the east considerably. Meanwhile, of course, the earth has revolved through only 1° around the sun.

This should be enough to show that Copernicus could predict an eastward drift of the moon against the stars as effectively as Ptolemy could explain why the drift of the moon is greater than that of the sun. Ptolemy's explanation is simpler, however, because only two motions are involved: a daily cycle of the moon around a fixed earth and a monthly cycle of the moon in a deferent. It is as if the moon moves 13° within its sphere as the sphere itself turns 360° a day. Copernicus, like Ptolemy, found it necessary to use a minor epicycle for the moon to reconcile differences between predicted and observed positions of the moon.

As with the moon, there are three simultaneous motions involved in understanding the movement of the planets: the earth's own rotation, its revolution about the sun, and the revolution of the planets. The basic explanation in Copernican theory for their apparent eastward drift against the background of the zodiacal belt is the same as for the sun and moon.

The earth and any observed planet are both revolving as the earth rotates. Therefore, when the earth completes a daily rotation, the planet is seen against a changed star background. Thus, for example, Mars will "rise" with a different star from one evening to the next, generally later according to its rate of revolution. Perhaps it would be preferable to say that Mars will appear above the horizon with a different star, rather than "rise" to emphasize the fact of the moving horizon.

In regard to retrograde motion, as with other observations, the earth is the frame of reference of the observer, and the zodiacal belt is the background frame of reference of the planet. The line of sight changes against the background of stars between any planet and the earth as both revolve around the

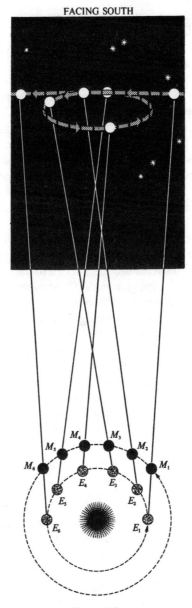

Figure 6.2

Retrograde motion of Mars. The Copernican explanation is that the earth revolves more rapidly than Mars. Each time the earth "catches up" to Mars (just after the E_3M_3 line of sight), Mars appears to reverse its eastward drift in the zodiacal belt of stars.

sun. Where an earthly observer sees a planet in the zodiacal belt depends upon this line of sight. For an illustration, see Fig. 6.3, which shows the retrograde motion of Mars. This planet, being farther than the earth from the sun, therefore moves more slowly around the sun than the earth does. Mars takes almost two years instead of one for a revolution. Notice that as the earth moves from E_1 to E_2 in the diagram, Mars circles from M_1 to M_2. The line of sight between them extended to the fixed background of stars shows Mars to be in an eastward drift with respect to the stars. Next, as the earth approaches E_3 in opposition to Mars, which is at M_3, the line of sight gives Mars the appearance of reversing direction toward the west with respect to the background of stars. In swinging from M_1 to M_3, Mars, of course, has not actually changed its uniform circular motion. By the time the earth and Mars have reached M_5 and E_5, respectively, the path of Mars reverts to the original eastward drift.

In the Ptolemaic theory, with the planets revolving about a stationary earth, epicycles were necessary to explain an apparent reversed motion. Then other epicycles upon epicycles had to be added to the original ones to account for other irregularities. At first glance, the Copernican answer is much simpler. Epicycles are not necessary for the original explanation of the observed reversed motions of planets. A changing line of light with a revolving earth gives the basic answer. But much of the simplicity was lost when Copernicus eventually found that he had to use many minor epicycles as well as eccentrics for adjustments between observed positions of planets and predicted positions. These adjustments could not be avoided as long as he used perfect circles and constant speeds for the orbits of all the planets, including the earth. But to use anything but circles and constant speeds would have been a sacrilege in terms of the doctrine of perfection of the heavens; Copernicus did not consider using anything else.

ON CHOOSING A MODEL

So far we have listed four basic motions observed in the heavens and described in detail just how these motions were explained by two contending conceptual models, the Ptolemaic and the original Copernican. If we could project ourselves back to the sixteenth century without knowledge of what has happened since, and if we were scientifically seeking a model of the universe, which model would we select and why? Such historical projection and analysis can give us an idea of what science is about.

Ideas as well as living things struggle for survival. This fact applies as well to major ideas within science. Seven criteria have evolved for evaluating a theoretical system in physical science. These criteria[2] are: (1) *explanation*, (2) *pre-*

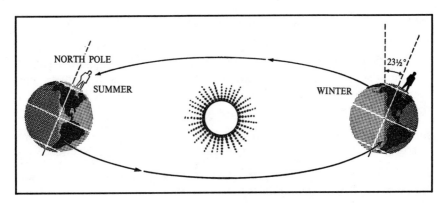

Figure 6.3

Seasons. The earth's axis is inclined at 23½° toward the sun in summer and 23½° away in winter. With the sun's direct rays alternately shifting north and south of the equator, an Arctic Circle observer has 24 hours of daylight on June 22 and 24 hours of darkness on December 22. On the vernal and autumnal equinoxes, there is an equal number of hours of day and night everywhere on the earth.

diction, (3) *flexibility*, (4) *functionality*, (5) *simplicity*, (6) *plausibility*, and (7) *falsifiability*. Let us briefly discuss these criteria and apply them to the Ptolemaic-Copernican issue.

1. Explanation. An effective theoretical model systematizes and explains. In other words, it unifies the data gathered by observation into a single system and gives the data coherence and consistency. It structures observation and explains it.

2. Prediction. This criterion primarily involves future data rather than past or present. A good model anticipates future events accurately and reliably. It is thus able not only to predict but also to be tested by its own predictions.

3. Flexibility. A theoretical system must have the means within itself of being reconciled with or adjusted to data; it must be modifiable. For example, in the case of the Ptolemaic and original Copernican systems, the geometric devices of eccentrics and epicycles provided such flexibility. Flexibility means being open-ended, as well as modifiable. A theoretical system must have plumules of growth from the known to the unknown. In modern terms, it must have lines to new research and to its own further development.

4. Functionality. Functionality is a vital criterion. A theory thrives on practical application; it becomes entrenched through technologies that it helps establish. Theoretical science and technology reciprocally nourish each other in modern research and development.

5. Simplicity. Science strongly emphasizes simplicity. This has been dignified into the *principle of parsimony,* which in effect states that, other things being the same, the simplest idea is the best. The simplest theory generally is the one that rests on the fewest basic assumptions.

6. Plausibility. Plausibility as a criterion refers to the extent to which a system of ideas fits in with other current systems of thought and experience.

7. Falsifiability. Karl Popper[3] in recent years has emphasized a criterion of "falsifiability." With flexibility, a good theory or model is modifiable, but it should not be endlessly so. There should be a clear basis in experience for rejecting as well as accepting a system of ideas.

The Ptolemaic and Copernican models during the sixteenth and early seventeenth centuries were contending systems that were able to account for observations with equal unity, coherence, and logic if one granted their assumptions. They were also able to predict with about equal accuracy. If you are still in doubt, compare again the Ptolemaic and Copernican explanation of the basic observations described earlier. Henri Poincaré (1854–1912) generalized that "If a phenomenon is susceptible of one mechanical explanation, it is susceptible of an infinity of others which would account equally well for all features revealed by experience." The word "infinity" is quite strong, but we can agree that for a given period of time, more than one good explanation is possible for the same set of phenomena. If, with Galileo and Newton, we become Copernicans, that is a real step forward from a seventeenth-century perspective. Hindsight is always easier than advance knowledge. Besides, the theory of relativity today holds that a modified Ptolemaic theory is as correct for an earthly observer as a Copernican theory for an astronomer elsewhere in the solar system.

With respect to the third criterion, both systems, as already noted, had great flexibility through their use of eccentrics, epicycles, epicycles upon epicycles, and so forth. Both could be adjusted to fit past observations and to predict future ones by using these devices and varying their radii, the value of the constant speeds used, and so on. The flexibility that Copernicus lost by abandoning equants he regained by adopting a moving earth.

The Ptolemaic system in the Western world was able to stimulate Arabian astronomers, medieval scholastics, and Copernicus to further observation, thought, and application. The original Copernican model led to one of the great intellectual and scientific revolutions of all time and stimulated such great trailblazers as Leonard Digges, Giordano Bruno, Johann Kepler, Galileo, and Sir Isaac Newton. Both models, therefore, had dynamic, flexible power to lead to further developments.

With respect to functionality, the Copernican system became comparable to the Ptolemaic. Shortly after the death of Copernicus, the mathematician Erasmus Reinhold used the added observations and improved numerical values of Copernicus's *Revolutions* to compute new astronomical tables called *Prulenic tables*. Both the book and the tables were used in the Gregorian revision of the calendar in 1582, but they have not even today totally displaced Ptolemaic tables for navigation.

This reduces the Ptolemaic-Copernican issue pretty much to a matter of plausibility, simplicity, and falsifiability. The Ptolemaic system had the advantage of plausibility. Appearances, common sense, Aristotle, philosophy, and religion all were against Copernicus. The preface to *The Revolutions,* written by Copernicus's friend Andreas Osiander, introduces the basic idea of the system, a moving earth, as one of mathematical simplicity and convenience. Copernicus may have been unaware of the approach used in the preface; it had been written for plausibility and acceptance. Mathematical or not, fear soon arose that this scheme would be interpreted literally, as Digges, Bruno, Kepler, and Galileo did in fairly short order, and that this interpretation would undermine the Bible, the Church, and established thought. Socioreligious repercussions resulted. The Inquisition, the new Protestant Church, and some Jewish communities lined up against it, as well as most astronomers: An earth-centered universe had Aristotle's physics to support it; Copernicus himself had no substitute system of mechanics to explain why the atmosphere moves with the earth, or why the earth does not fly apart, and so on.

Plausibility was thus against Copernicus: a sun-centered universe did not fit in with established ideas of the time. That did not, however, make Copernicus subversive or "wrong." Nor, on the other hand, does it mean that all new, bold ideas have what it takes to survive. Plausibility is a difficult criterion. Who can judge what will be plausible tomorrow?

Since plausibility can change with time and with the development of knowledge and experience, we are left with only the criteria of simplicity and falsifiability. Is Copernicus's system simpler? We have seen that, in principle, it is: A single earth rotating daily is simpler than having the sun, the planets, and all the stars making daily circles around the earth. A revolving earth eliminates major epicycles from the retrograde motion of the planets. Elimination of equants is also of value for simplicity.

Although seemingly much simpler an principle, the original Copernican system is actually not so simple when worked out in detail. In spite of introducing a moving earth into his scheme, Copernicus had to add ever more and more to the 34 minor epicycles and other circles he used in his early estimates for his system. Otherwise, planets would not show up at the predicted time or place. Copernicus, therefore, never fully realized his original hope of the advantage of simplicity in his system. But what if the universe is not a simple one? Then the best theories would not necessarily be the simplest. In either case, the Ptolemaic and Copernican models would still be about equal contenders for simplicity.

Neither the Ptolemaic nor the original Copernican system, however, could measure up to the standard of falsifiability. Each system could endlessly modify itself with epicycles and eccentrics to correspond to data. True, each system became cumbersome with these geometric devices, but that is the point: No observational test could reject the system; the devices ingeniously prevented that. Besides, the universe itself could be complex.

Copernicus's *Revolutions* nonetheless quite boldly robbed the earth of its unique status as the center of the universe. This had tremendous psychological, religious, philosophical, and astronomical significance—at least enough to challenge the even bolder spirits of Digges and Bruno and, ultimately, to spearhead the Scientific Revolution.

The Ptolemaic-Copernican issue could not remain deadlocked. Yet Copernican boldness alone was insufficient. Copernicus left too many questions unanswered, such as undetected stellar parallax, an atmosphere moving with the earth, and the earth remaining intact at great speeds. There were also such questions as these: How can a doctrine of perfection of the heavens be reconciled with an imperfect, heavy earth moving around the sun together with supposedly perfect, weightless planets? And from where do the necessary forces arise to push the heavy earth along with other planets around the sun? Such questions will be answered in later chapters.

FAITHS AND MODELS

In summary, astronomy was characterized early by guiding faiths in a mathematical order, unity, and simplicity in nature. The original aesthetic, philosophic-religious character of these faiths are expressed in such concepts as Pythagorean "music of the spheres" and Platonic "perfection of the heavens." Observed periodic motions of celestial bodies lent themselves to geometric treatment. Encapsulating celestial spheres were projected upon the heavens. Reference frames and an elemental relativity of motion were recognized implicitly at least. Geometry shaped astronomy in the form of conceptual models whether Ptolemaic or Copernican. Such geometric models enabled astronomical prediction, as well as modification of models where observed sightings of celestial objects did not match predicted sightings in time and place. Such scientific operations had all the earmarks of modern science: quite similar guiding principles leading to mathematical models that successfully predicted events. In searching for order, astronomers were successfully assuming order of particular molds.

Science is thus more fully human than detached reasoning from data. Aesthetic, intuitive, subjective qualities are involved in the guiding faiths or principles that mold specific conceptual models and their specific assumptions. Interaction—not detachment—is also involved in observation. We saw that astronomers created and projected celestial spheres that frame moving objects and partly determine how they are seen and described. We also saw that *relative* motion between observer and observed affects observation. Thus, conceptual models, through the guiding principles that shape them and their application to nature, reflect the more fully human and interactive character of science.

NOTES

1. Unknown to the West, the printing press had originated in China thirteen centuries earlier.

2. For the first six criteria, we credit Gerald Holton and Duane Roller, *Foundations of Physical Science* (Reading, Mass.: Addison-Wesley, 1959), as well as similar ideas of other science historians.

3. For the criterion of falsifiability, see Karl Popper's *Conjectures and Refutations* (New York: Basic Books, 1962), particularly chapter 1.

SUGGESTIONS FOR FURTHER READING

Bernal, J. D. *Science in History.* Cambridge, Mass.: MIT Press, 1971, vol. 2, ch. 7: "The Scientific Revolution."

Dreyer, J. L. E. *A History of Astronomy from Thales to Kepler.* New York: Dover, ch. 13: "Copernicus."

Kuhn, Thomas S. *The Copernican Revolution.* New York: Vintage, 1959.

Pennekoek, A. *History of Astronomy.* New York: Interscience, 1961, part 2: "Astronomy in Revolution," chs. 16–19.

7

The Book of Nature:
Brahe's Data, Kepler's Laws, and Galileo's Telescope

Science is written in this great book, the Universe,
which stands continually open to our gaze.
                    Galileo, *Dialogues Concerning Two Chief World Systems,* 1623

Go to the horse's mouth to count the teeth.
                    Francis Bacon, *Novum Organum,* 1620

BRAHE'S PASSION FOR PRECISION

Let more precision decide! In effect, that characterizes the attitude of the great
Danish astronomer Tycho Brahe (1546–1601) toward the problem of celestial
motions.

At the age of fourteen, Brahe was greatly impressed by the eclipse of
August 21, 1560. It was not the eclipse alone that fascinated him, but the fact
that it had been predicted. In succeeding years, a wealthy uncle sponsored him
in the study of law, but he avidly read astronomy books that he purchased
secretly. In November 1572, a new star, or *nova,* suddenly appeared. It was this
nova's explosion into view that fired Brahe into a lifetime of astronomical
observation. The new star, almost as bright as Venus, was visible for about
eighteen months before disappearing. Brahe meanwhile observed all changes
in its brilliance, measured angles of its position, and proved to his own satis-
faction that the newcomer was as distant as the fixed stars. What was a new
comer, and a temporary one at that, doing in a region supposedly as permanent
and unchanging as the celestial sphere of stars? Then, in 1577, a new comet
appeared and set Brahe further aflame. The comet's path was clearly through
a number of the impenetrable crystal spheres that supposedly made up the
machinery of the heavens. How was this possible?

99

Enthusiastic as well as able, Brahe by this time (1577) was famous for his astronomical observations and lectures. Frederick II, King of Denmark, placed the Island of Huen, large funds, and a pension at his disposal. On Huen, Brahe built the most efficient observatory yet known, Uraniborg. Most important of all to Brahe were the huge sighting instruments that he had been able to construct in his observatory. (The telescope had not been invented yet.) With a passion for precision, he gave himself the most accurate naked-eye instruments in history.

For about twenty-one years at Uraniborg, Brahe pinpointed the positions of about eight hundred stars. More than that, he traced the positions of the planets entirely through their courses rather than at selected points. He systematically recorded these observations with an unheard-of accuracy into books of tables later famous as the *Rudolphine tables*. Copernicus had operated with an accuracy of about 10', or $1/6°$, of arc.* Brahe's precision was twenty times better, with about 0.5', or $1/120°$, of arc. He even corrected for light-bending, or refractive effects of the atmosphere. For Brahe, it was improved instruments and techniques, more precise observation, and systematic plotting of data that would establish the pattern of the universe. His genius was that of infinite patience for systematic observation.

Among other things, Brahe carefully looked for stellar parallax. If the earth revolves, closer stars should be seen shifting yearly against backgrounds of more distant stars. But even with the most precise instruments of the time, Brahe did not find the parallax he sought. This confirmed his Ptolemaic leanings. Not satisfied, however, with the original Ptolemaic model, Brahe developed a scheme of his own. Into his system, known as *Tychonic,* the planets revolved around the sun—but the sun with all its attendants revolved around a stationary earth. Clearly, Brahe was moving toward the Copernican theory—and stopped short: since greater precision revealed no stellar parallax,[1] the earth must be stationary.

In addition to his infinite patience for precision and detail, Brahe also had an independent nature, a fiery temper, and arrogant ways. These earned for him powerful enemies at court. When his friend King Frederick died in about 1592, those enemies were able, within a mere five years, to force him from his island, his work, and even his country. Huen ceased to exist as an observatory.

After two years of wandering, Brahe was invited by Rudolph II, Emperor of Bohemia, to settle with his instruments in Prague. There he again set up an observatory in a castle and continued his celestial explorations and lectures. In 1601, however, broken in health, he died.

---

*I.e., a small portion of a 360° circle.

KEPLER, MATHEMATICAL MYSTIC

A year or so before his death, Brahe hired a young German assistant, Johannes Kepler (1571–1630). As an astronomer, Kepler already possessed exceptional mathematical ability and imagination. Brahe had recognized this in Kepler's first book in astronomy, *Mysterium Cosmographicum (Cosmographic Mystery)* (1596), and invited Kepler to join him in Prague. Because of serious religious turmoil in his native German province of Styria, Kepler accepted Brahe's invitation. This was fortunate indeed because when Brahe died a year later, his precious data—the largest, most accurate astronomical records up to that time—could not have fallen into better hands.

Genius takes many forms. Kepler, a confirmed Copernican, was a mathematical mystic with great respect for Brahe's data. If Brahe's great talent was that of the observer with infinite patience for acquiring data, Kepler's was that of the theoretician with a bold mathematical imagination for persistently, even fanatically, ordering the data.

Kepler was born in Germany on December 27, 1571, of parents of fallen, lesser nobility. From early childhood his personal life was one of persistent sickness, poverty, and frustration. It was in the planets and stars that he found most of his life's consolation.

Quite early, Kepler, like the ancient Pythagoreans, became convinced that geometry and numbers are the essence of the universe. As a young lecturer at the university of Graz, he was consumed with the idea of connecting the five regular solids of geometry—the tetrahedron, cube, octahedron, dodecahedron, and icosahedron (Fig. 7.1)—with the orbits of the six known planets. In applying these solids to the solar system, Kepler arranged them in succession one within the other, as described in *Mysterium Cosmographicum.* In Kepler's plan, the planetary orbits were circles on spheres placed within and around the geometric solids.

But great men are not always successful in their conceptual schemes, in science or otherwise. No matter how hard Kepler tried, Brahe's meticulous data would not support perfectly circular planetary orbits. The geometric solids eventually failed as a model for explaining the orbits' size, number, and arrangement.

Before Brahe died, Kepler promised him that he would do what he could to test Brahe's own model with the data. At first, Kepler faithfully tried to reconcile his colleague's model with the evidence. But unsuccessful in this, Kepler then tried the original Copernican model. That model would not line up with the data either, except approximately, and with an ever increasing complexity of epicycles, eccentrics, and even equants. Through the use of an eccentric circular orbit, Mars came within 8' of arc, or $1/7°$, of its predicted position. But Kepler, having too much confidence in Brahe's precision to question his data, rejected the model. If circular models did not closely enough conform with the data, then away with circles. In that case, the pattern of planetary orbits would have to appear directly from the data itself, much like the pattern that appears when a child traces a line from dot to dot in a pictorial puzzle. For the first time in modern history, conceptual schemes were rejected on the basis of precision of measurement. But Kepler had fanatical faith that when the dots were correctly connected, a mathematically simple universe would appear. Finally, after eighteen years of exhausting labor, he emerged with his three laws of planetary motion. These laws did simplify the Copernican theory—amazingly so! And they provided foundations for Isaac Newton's amazing law of universal gravitation.

KEPLER'S CELESTIAL HARMONIES

Kepler found that once the planets were freed from circular orbits, their plotted positions formed ellipses. First discovering this with Mars, he was delighted to find the same orbital pattern for other planets. Best of all, the elliptical orbits appeared only when he *assumed* his frame of reference, the earth,

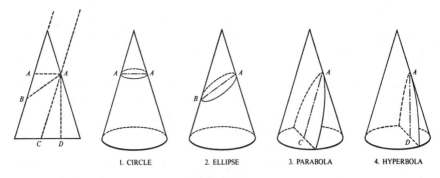

1. CIRCLE   2. ELLIPSE   3. PARABOLA   4. HYPERBOLA

Figure 7.1

Conic sections:

1. Circle: A section parallel to the conebase.
2. Ellipse: a section not parallel to the base.
3. Parabola: a section parallel to the cone side.
4. Hyperbola: a section beyond the parallel to the side.

to be revolving around the sun as well. Now he needed no celestial spheres, no circular deferents, no epicycles upon epicycles, no further geometric devices. After the elimination of the single assumption of perfection of the heavens, all planetary motions fell into line, the line or pattern of an ellipse if the earth itself was equated with the "wanderers" in the night sky. All this and more is implied in Kepler's first law of planetary motion: *Every planet moves in an elliptical path with the sun at one focus.*

But just what is an ellipse? A heritage of ancient geometry is the mathematical model of conic sections. Fig. 7.2 shows that an ellipse is one of four sections formed by properly cutting a cone. When a circular cone is cut from point A by a plane parallel to its base, the cross section is a circle. If the cone is cut from A at an angle to the base, the new section forms an ellipse. Thus, a circle is a special, limiting case of an ellipse that depends upon the angle at which the cone is cut at point A.

It follows from Kepler's first law that a planet in its elliptical orbit will be at varying distances from the sun at different positions. The point at which a planet is closest to the sun is called the *perihelion.* The furthest point is called the *aphelion.* Kepler found that the elliptical orbits of the various planets are only slightly elongated, that is, almost circular. The sun is close in each case to the symmetrical center, and the distances between it and the aphelion and perihelion of each planet are fairly close. It was shown above that a circle is a limiting form of an ellipse (see Fig. 7.1). Astronomers before Kepler, including Copernicus, had been restricting themselves to a limiting form of the ellipse— at a great sacrifice of simplicity. Assumptions can be encapsulating indeed!

Although the first law was beautiful in its simplicity, it was insufficient. Alone, it did not provide for the speed of motion of the planets. Accurate observations showed that the speed of a planet in its orbit varies. Is there a regular basis for this variation? If so, what? Without knowledge of how planets change their speeds in elliptical orbits, no reliable predictions could be made of their future positions. Thus the ellipse in all its simplicity would not be able to displace the former Ptolemaic or Copernican models. These earlier models were cumbersome, but they at least could make useful predictions of planetary positions by assuming constant speeds.

Kepler was confident of the mathematical simplicity and uniformity of nature. He had previously used equal triangles as a mathematical tool in handling areas of circular orbits; he now applied this tool to elliptical orbits. The result, after nine more years, was the second law, the law of equal areas, first described with the first law in his book *Astronomia Nova (New Astronomy)* (1609).

In the second law, the mathematical formulation is the same for all planets. *The law states simply that an imaginary line drawn from any planet to the sun sweeps over equal areas in equal times.* For this statement to be true, the closer a planet is to the sun, the faster it has to move (Fig. 7.2).

Once "perfectly circular" orbits for planets were gone, so was another old assumption, a constant orbital speed. It also follows that a planet's speed is greatest when closest to the sun (perihelion) and slowest when furthest from the sun (aphelion). (A child knows that the shorter the length of a cut from a pie, the wider should be the crust to obtain an equal share.) Kepler discovered that with a high degree of accuracy for a planet in a nearly circular orbit, the following relationship holds: $vd = k$ (a constant), where v is the velocity of a planet at any point in its orbit and d is its distance to the sun at that point. That is, the product of velocity v, and the distance d of the planet at various points in its orbit approximately remain the same. Kepler had the joy not only of showing a reasonably accurate mathematical relationship between planetary positions and speeds in elliptical orbits, but also of finding the relationship confirmed by Brahe's data.

The second law of planetary motion implies a simple mathematical pattern of change of motion by a given planet in its own orbit. As such, it enables prediction of the position of a planet at any time. It does not, however, indicate a relationship between the motions of different planets. Kepler's zeal for a basic mathematical unity of the solar system drove him on to seek this larger relationship. For another nine years he labored.

And then he found what he was after in the form of an equation that joins the sun, earth, and other planets mathematically into an interrelated system of motion apart from the rest of the universe. Kepler expresses his own exaltation at the discovery of his third law of planetary motion, the harmonic law, in his great work *Harmonice Mundi* (*On Celestial Harmonies*) (1618):

> It is not eighteen months since I got the first glimpse of light, three months since the dawn, very few days since the unveiled sun, most admirable to gaze upon, burst upon me. Nothing holds me; I will indulge my sacred fury; I will triumph over mankind by the honest confession that I have stolen the golden vases of the Egyptians to build up a tabernacle for my God far away from the confines of Egypt. If you forgive me, I rejoice; if you are angry, I can bear it; the die is cast, the book is written, to be read either now or by posterity, I care not which; it may well wait a century for a reader, as God has waited six thousand years for a observer.[1]

Here, as he first states it, is Kepler's discovery:

> And so if any one take the period say of the Earth, which is one year, and the period of Saturn, which is thirty years, and extract the cube roots of this ratio and square the ensuing ratio by squaring the cube roots, he will have as his numerical products the most just ration of the distances of the Earth and Saturn from the sun. For the cube root of 1 is 1, and the square of it is 1; and the cube root of 30 is greater than 3, and therefore the square of it is greater than

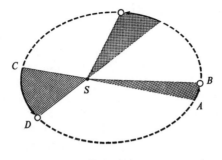

Figure 7.2

Law of equal areas. An imaginary line from any planet to the sun sweeps over equal areas in equal times area *ASB* = area *CSD*.

9. And Saturn at its mean distance from the sun is slightly higher than nine times the mean distance of the Earth from the sun.[2]

The law stated simply is this: All planets have the same ratio of the square of time of revolution around the sun (T^2) to the cubes of average distance from the sun ($\bar{D}^3$). In algebraic symbols, this reads as

$$\frac{T^2}{\bar{D}^3} = k$$

(k is a constant value), regardless of which planet we choose.

To really see what so excited Kepler, complete columns 4 and 5 in Table 7.1. Note that in column 6 the ratio

$$\frac{T^2}{\bar{D}^3} = k$$

is the same for each planet. The average distance of $\bar{D}$ of any planet can be calculated by knowing just its period of revolution T. The earth's distance to the sun is 93 million miles or *1au*, by definition. When years are used for T and astronomical units (*au*) are used for $\bar{D}$, then the ratio

$$\frac{T^2}{\bar{D}^3} = k$$

is 1, or $T^2 = \bar{D}^3$. More than that, Kepler's harmonic law unites all planets and satellites into one solar system of motion. Outside this system are stationary stars widely spread in a thick celestial sphere. In a sense, with Kepler, the single, sun-centered world system of Copernicus became separated into a solar family and the surrounding stars. The solar system became an entity of its own.

FAITHS, ASSUMPTIONS, AND HYPOTHESIS TESTING

The work of Kepler highlights the role of assumptions in science and the necessity for constantly examining the assumptions behind *all* that we think or do. Copernicus had kept his own hands tied by the assumptions of uniform and circular motion and by a faith in the perfection of the heavens behind these assumptions. Kepler, Pythagorean mystic that he was, also felt deeply about circular orbits. We have seen that he did everything he could with circles, first to make Brahe's scheme work, and then to demonstrate Copernicus's model. When circles became too cumbersome to suit his aesthetic sense as well as his respect for accuracy, Kepler gave up the circles. He abandoned faith in the perfection of the heavens for simplicity. With the ellipse, Kepler, in one blow, cut a highly entangled though useful Gordian knot, and the solar system became simple. There was no need for a cumbersome system of deferents, epicycles, or equants—only for elliptical orbits tied together in the simple, unifying interrelating equality of ratios:

$$\frac{Te^2}{De^3} = \frac{Tx^2}{Dx^3} = \ldots = 1,$$

where the first ratio is that for the earth and the second ratio is that for any other planet, if the proper units are used as in Table 7.1. A significant simplicity indeed! And one that justified Kepler's cherished Copernican assumption that the earth and all the planets move around the sun.

Kepler's law illustrate the operating scientific principle (really faith) of parsimony: the simplest theory is the best, other things being the same. To this day, scientists seek the simplest theories to unify and explain what they see.

Table 7.1
Harmonic Law Relationships

1 Planet	2 Time of one Revolution $T(YR)$	3 Average Distance to Sun $D\,(AU)$	4 T^2	5 $\bar{D}^3$	6 $\dfrac{T^2}{D^3}$
Saturn	29.46	9.54			1
Jupiter	11.80	5.20	140.66	140.61	1
Mars	1.88	1.52			1
Earth	1.00	1.00	1.00	1.00	1
Venus	.62	.72			1
Mercury	.24	.39			1

And certainly the simplest are the most convenient. But what if the universe itself is complex and its phenomena are therefore more truthfully portrayed by a complex theory? Should convenience be preferred to truth in that case?

Agreed, Kepler in his elliptical model removed small errors of prediction obtained with Ptolemaic or original Copernican models. Ptolemaic theorists, however, by adding still more epicycles to their already cumbersome model, could have approached even Kepler's accuracy. Even better, what if Kepler had been a Ptolemaic and had substituted ellipses or other appropriate forms for circles in Ptolemaic theory? The Ptolemaic model would have been simpler, too.

If any three words describe modern scientific methodology, they are "mathematical hypothesis testing." Kepler, in this writer's opinion, along with Galileo and Newton, was an initiator of modern hypothesis testing. Regardless of his strong personal preferences, we have seen that he tested a number of hypotheses on the same problem of planetary orbits: Platonic solids, Brahe's geocentric circles, Copernican circles, the ellipse. In his hands, these ideas became hypotheses that were accepted or rejected on the basis of Brahe's data. It was the data that decided. With Kepler, ideas were tested by careful observation; data were not to be forced or "saved" by ideas. For example, if the data on Mars led Kepler to reject circles and suggested to him an ellipse, then the ellipse hypothesis had to be tested and retested by data, first for Mars and then for other planets.

The ideas tested by Kepler in his second and third laws are algebraic. For example, by considering the equal areas swept over by a planet in the second law, he produced an equation relating quite closely the positions and speeds of a planet: $v_1 d_1 \approx v_2 d_2 \ldots \approx$ a constant k for that planet. Observation showed that planetary positions could be predicted from this equation with considerable accuracy. The equation $T^2 = k\overline{D}^3$ also symbolizes Kepler's harmonic law. With Kepler, equations were displacing epicycles. Algebraic laws for the first time were displacing geometric ones in physical science. Actually, Kepler was operating algebraically without using algebraic symbols as shorthand, as can be seen in his first statement of his harmonic law, quoted earlier. Kepler, in initiating such a new mathematical form of law, was soon joined by his great contemporary Galileo, operating independently in mechanics. The equation as a form of law came into full bloom in Newton's laws of motion and gravitation, and after that, more and more laws took the form of an equation in the physical sciences.

Guiding principles or faiths are subjective factors, indispensable for model building, discovery, and hypothesis testing in science, as the work of Kepler certainly shows. But they are valid only in correspondence with the book of nature, not in place of it. When Kepler rejected circular orbits for Mars, he had the courage to choose Brahe's precise data and his own faith in simplicity in preference to the Platonic faith in "perfection of the heavens." Likewise, as we shall see, Galileo chose his telescope and the book of nature in preference to Aristotelian doctrines.

PTOLEMY, COPERNICUS, AND GALILEO'S TELESCOPE: THE HEAVENS BROUGHT CLOSER

Galileo Galilei (1564–1642) was a Copernican colleague of Kepler. Although the two never met, they encouraged each other through correspondence. Galileo was born on February 18, 1564, in Pisa, Italy. The famous Leaning Tower of Pisa is still associated with his name. While there is no evidence that Galileo actually dropped balls from the top of the tower in public demonstrations, it has become symbolic of Galileo's work with falling objects and of his Copernican convictions.

Galileo's father, Vincenzio, a descendant of a long line of Florentine nobility, had moved to Pisa to repair family finances. A cultured man and a musical rebel, he and a dedicated group of friends, endeavoring to replace older musical forms with new ones, invented the recitative and other early operatic forms. Galileo, tutored as a boy by his father, benefited early from his father's love of music, books, and mathematics. Like his father, the son could draw artistically and had exceptional literary talent. The young Galileo brought an innovating spirit to astronomy and mechanics as well as to the arts.

Vincenzio sent his son at the age of seventeen to the University of Pisa to study medicine. But the mathematics of Euclid and Archimedes proved to be of greater fascination to Galileo than the ancient medical theories of Aristotle and Galen. In any case, his supposed observation of a chandelier swinging back and forth during a sermon in the Cathedral of Pisa led to Galileo's invention of a device to count pulse beats, and cut short his medical studies.

In 1600 the controversial Giordano Bruno was burned alive at the stake by the Inquisition for Copernican and other religious heresies that he would not recant. At the time Galileo was Lecturer of Mathematics at the University of Padua. In 1604, when another nova spectacularly appeared in the heavens, Bruno's fate did not prevent Galileo, in the freer atmosphere of Padua, from lecturing on the Copernican implications of this phenomenon. He remained comparatively quiet as a confirmed Copernican, however, for about another five years. Galileo expressed his caution at this time in a 1610 letter to Kepler:

> I have written many arguments in support of him [Copernicus] and in refutation of the opposite view—which, however, so far I have not dared to bring into the public light, frightened by the fate of Copernicus himself, our teacher, who, though he acquired immortal fame with some, is yet to an infi-

nite multitude of others (for such is the number of fools) an object of ridicule and derision. I would certainly dare to publish my reflections at once if more people like you existed; as they don't, I shall refrain from doing so.

In 1609, a rumor reached Galileo that a Dutch lens grinder, Lippershey, had devised an instrument with two lenses that brought distant terrestrial objects into closer view. That was all Galileo needed as a clue. Within twenty-four hours he had constructed the first of several telescopes that he excitedly turned to the skies. For Galileo, the new instrument was an extension of the senses: through it, the heavens could be observed with a detail never before possible. Direct evidence was at hand to put the Ptolemaic-Copernican controversy to a final test. Galileo was hopeful that his colleagues would be willing to observe and to rethink the issue.

And observe Galileo certainly did! What that stargazer found he openly reported in his book *Sidereus Nuncius* (*The Starry Messenger*), published in 1610. The following are some of the significant observations and inferences of this work:

1. There were mountains, craters, and plains on the moon—much like on the earth. How could the moon be irregular in surface and still be a perfect sphere with a smooth, crystalline face?

2. There were dark spots moving across the face of the sun that apparently rotated around it in about twenty-seven days. What were large spots doing on a celestial object composed of an unchanging, "incorruptible" substance?

3. There was a bulge around Saturn that prevented perfect sphericity for that body.

(An implication of these first three observations, for Galileo, was that the above celestial bodies were imperfect in form and earthlike in substance. The earth, therefore, could take a place in space with other objects moving around the sun.)

4. Even more important an inference was the sight of four moons revolving around Jupiter. This, for the first time, provided direct evidence of heavenly bodies moving directly around another celestial body instead of directly around the earth. If such objects could move around Jupiter, then why couldn't other objects (planets) revolve directly around the sun?

5. Venus had phases, as predicted by Copernicus. That is, Venus completed a series of phases from new to full to new, like those of the moon. Observed phases for Venus meant that the planet did not shine of its own light, as formerly expected of a perfect heavenly body. Like the earth, Venus shone merely with reflected light. Here again, the same standard was applied to earth as to other celestial bodies.

There was also about a 5:1 ratio in the apparent diameters of Venus when in its narrow crescent and full phases. This ratio most likely meant that Venus was about five times closer to the earth in its new phase than in its full phase. This could be easily explained with both the earth and Venus revolving around the sun in the Copernican model. The ratio was explainable through Ptolemaic epicycles albeit with great difficulty. But with Venus always between the sun and a stationary earth (see Fig. 5.4), the original Ptolemaic model certainly could not explain the nearly full phase seen for Venus. Galileo believed that the phases of Venus were decisive evidence against the Ptolemaic theory.

6. There were myriads of stars in the Milky Way not previously seen. And yet even with Galileo's telescopes of up to 30 power, no star was ever more than a pinpoint of light—nor are they today, with incomparable more powerful telescopes. Distances to the stars were therefore much greater than formerly supposed. Such immense distances supported Copernicus's original contention that stellar parallax had not been detected from a revolving earth because stars are too far away.

Galileo's findings scientifically weakened some of the props of an earth-centered system. These findings, however, for reasons given below, were not quite so crucial as he thought:

1. Mountains on the moon, sunspots, and an asymmetrical Saturn seriously undermined the concept of perfection of the heavens, but the earth scientifically could still be the center of a world system without the philosophic-religious assumptions of perfect heavenly bodies, perfect spheres, or perfect circles. That is, the Ptolemaic theory could be reestablished on other assumptions, just as the Copernican theory had been reinstated with Kepler's ellipses.

2. The existence of moons revolving around Jupiter does refute the claim that the earth is the *direct* center of revolution of *all* objects in the universe, but it does not eliminate the possibility that Jupiter itself with its satellites could be revolving around the earth.

3. The Ptolemaic model, as it is customarily understood, is not able to account for the phases of Venus. Ptolemaic theory can get around this, however, by having Mercury and Venus revolve around the sun as the sun rotates around the earth (Heraclides, fourth century B.C.E.) or by having all planets revolve around the sun as the latter moves around the earth (Brahe, sixteenth century C.E.). The latter modification can qualitatively explain complete phases seen for planets. Both of the alternatives, however, are admittedly compromises that seriously weaken Ptolemaic theory. Having all planets revolve around the sun as the latter encircles the earth means getting ever further away from the original assumption of the earth as the direct center of all

things. If all other planets revolve around the sun, it is but one more step for the earth to do the same. Brahe himself didn't take this last step to the Copernican theory because he looked for stellar parallax and couldn't find it. And yet the fact that Brahe's revision of the Ptolemaic model can be used to explain the phases of Venus means that Galileo's telescopic observations did not deal quite the final blow to an earth-centered system that Galileo thought they would.

GALILEO AND ACADEMIC FREEDOM

Instruments as extensions of the senses can settle theoretical disputes—if men are receptive to instruments. Galileo's telescopic appeal to the senses came at a time when appeal to authority rather than to direct evidence still prevailed among educated men. The senses were considered to give illusory, transitory details rather than basic, permanent principles and could not be trusted. Most of Galileo's colleagues therefore ignored his appeal, his observations, and his conclusions.

As an example of this, consider the following argument against the existence of Jupiter's satellites by Francesco Sizzi, a seventeenth-century Florentine astronomer:

> There are seven windows in the head, two nostrils, two eyes, two ears, and a mouth; so in the heavens there are two favorable stars, two unpropitious, two luminaries, and Mercury alone undecided and indifferent. From which and many other similar phenomena of nature, such as the seven metals, etc., which it were tedious to enumerate, we gather that the number of planets is necessarily seven.
>
> Moreover, the satellites are invisible to the naked eye, and therefore can have no influence on the earth, and therefore would be useless, and therefore do not exist.
>
> Besides, the Jews and other ancient nations as well as modern Europeans have adopted the division of the week into seven days, and have named them from the seven planets: now if we increase the number of the planets this whole system falls to the ground.

With respect to such arguments as the above, Galileo quite caustically replies in a letter to Kepler (1610):

> We will laugh at the extraordinary stupidity of the crowd, my Kepler. What do you say to the main philosophers of our school, who, with the stubbornness of vipers, never wanted to see the planets, the moon or the telescope although I offered a thousand times to show them the planets and the moon. Really, as some have shut their ears, these have shut their eyes toward the

light of truth. This is an awful thing, but it does not astonish me. This sort of person thinks that philosophy is a book like the *Aeneid* or *Odyssey* and that one has not to search for truth in the world of nature, but in the comparisons of texts (to use their own words).

The telescope and *Starry Messenger* won fame for Galileo beyond the boundaries of his own country. When an appreciative free Republic of Venice granted him life tenure in his professorship at the University of Padua and doubled his salary, Galileo took heart. Against the advice of friends, he left Padua in 1610, where he had been honored and sheltered for eighteen years, to return to Pisa as chief mathematician to the Grand Duke Tuscany and as head of mathematics at the University of Pisa. In 1611, Galileo was fêted in Rome and elected to the new Accademia dei Lincei, one of the earliest scientific societies.

Galileo, an exceptionally able polemicist, was confident that he could convince the highest ranks of the Church that the new astronomy did not contradict religious thought, that the "same Creator was behind both the Bible and Nature." He become more and more open in his Copernicanism. In 1615, Galileo was asked to come to Rome to defend the Copernican position before high Church officials. He had a number of friends in the Church hierarchy; many cardinals also maintained an attitude of compromise about the issue. There was, on the other hand, a strong ecclesiastical faction pressuring the pope and the College of Cardinals for a final, clear-cut decision on the heretical character of Copernicanism. Galileo's persuasive talents and friends proved to be insufficient. The Congregation of the Index ruled against the Copernican system. Copernicus's *Revolutions* was banned and placed on the *Index Expurgatorius (Index of Forbidden Books)*. Officially warned against teaching Copernicanism, Galileo returned home heartsick. For a number of years, he devoted himself to noncontroversial problems in hydrostatics.

When Cardinal Barberini was elected Pope Urban VIII in 1623, Galileo's Copernican heart beat faster, for the new pontiff was a personal friend of Galileo and an enlightened man. Galileo soon discussed with the pope plans for a book, *Dialogue on Systems of the Universe,* in which the Ptolemaic and the Copernican systems were each to be represented by a spokesman. A third participant in the *Dialogue* would be neutral and open. The pope acquiesced to an impartial representation of the two points of view. Printed nine years later, the *Dialogue* sold extensively for six months. Then suddenly the book was banned. An ecclesiastical faction hostile to Galileo convinced the Church that the *Dialogue* benefited the Copernican cause and that Galileo had circumvented the warning given him in 1616. The *Dialogue* remained on the Index list with Kepler's *Epitome of Copernican Astronomy* and Copernicus's *Revolutions* for over two hundred years.

For championing the Copernican over the Ptolemaic system, Galileo was called to face the Inquisition Court in 1633. Whatever the details of his three months' imprisonment, in June 1633, Galileo, on his knees before the court, officially recanted his belief that the "Sun is the center of the World and that the Earth moves around it."

For the remaining nine years of his life, Galileo worked quietly under house arrest at his farm on the outskirts of Florence. There, his great accomplishment—perhaps his greatest—was the further development of a new theory of mechanics. This theory was described in his work *Dialogues Concerning Two New Sciences.* A visiting Dutch publisher smuggled the manuscript out of Italy after its completion in 1636. The book was first published in Leyden in 1638.

Two centuries after Galileo's death, more powerful telescopes provided final evidence of a revolving earth. Ptolemaic astronomers for ages were insisting that stellar parallax would be observable if the earth revolved. Brahe, searching for but not finding parallax, stopped short of becoming a Copernican, as we have seen. The Copernican explanation that the stars are too distant for observation of parallax appeared defensive and certainly inconclusive. Then, in 1838, the German astronomer F. W. Bessel did find parallax. The stellar shift was slight but unquestionable, and it was followed by other cases. Astronomical instruments had finally become powerful enough to detect the yearly shifts of closer stars against the backgrounds of distant stars. Men reasoned brilliantly with mathematics on both sides of the question for thousands of years before the increased resolving power of a telescope cut through the entanglement. Men need everything they have and can invent—mind, imagination, and sense-extending apparatus—to come to grips with nature.

The following extract from a preface by Einstein to Galileo's *Dialogues* is appropriate at this point:

> The *leitmotif* which I recognize in Galileo's work is the passionate fight against any kind of dogma based on authority. Only experience and careful reflection are accepted by him as a criterion of truth. Nowadays it is hard for us to grasp how sinister and revolutionary such an attitude appeared at Galileo's time when merely to doubt the truth of opinions which had no basis but authority was considered a capital crime and punished accordingly. Actually we are by no means so far removed from such a situation even today as many of us would like to flatter ourselves; but in theory, at least, the principle of unbiased thought has won out and most people are willing to pay lip service to this principle.[3]

The crime in Galileo's case was not in the Church's maintaining Ptolemaic convictions held by most astronomers and churchmen—for geocentrism could still have been justified even scientifically at the time—but in its ruthless crushing of new ideas. Ideas are not necessarily better because they are new, but the way should be open for honest examination of them for their possibilities. Expanded literacy and progress necessitate that intellectual honesty be a common virtue.

Galileo's telescope today is a symbol. Pointed as it was to the heavens, it symbolized an appeal to the senses, to direct, organized observation of nature. In bringing the heavens closer to the earth, he was placing astronomy on a firmer empirical basis. A Scientific Revolution had begun. To catch hold of this revolution and to support and extend it to an Industrial Revolution, a rising middle class would soon make itself felt in Europe.

Although today the experimental testing of hypotheses has been established as indispensable to science, intellectual honesty and academic freedom are still contentious issues everywhere, often even in science. Sir Oliver Lodge in 1893 dramatically exclaimed:

> I have met educated persons who, while they might laugh at men who refused to look through a telescope lest they should learn something they did not like, yet also themselves commit the very same folly. . . . I am constrained to say this much: Take heed lest some prophet, after having excited your indignation at the follies and bigotry of a bygone generation, does not turn upon you with the sentence, *"Thou art the man!"*[4]

NOTES

1. Johannes Kepler, *Opera omnia,* ed. Frisebus (Frankfurt, 1853), p. 71.
2. Ibid.
3. In *Dialogues,* trans. Stillman Drake (Berkeley and Los Angeles: University of California Press, 1953).
4. Oliver Lodge, *Pioneers of Science* (New York: Dover, 1960).

SUGGESTIONS FOR FURTHER READING

Bernal, J. D. *Science in History.* Cambridge, Mass.: MIT Press, 1971, vol. 2, ch. 7: "The Scientific Revolution."
Cohen, I. Bernard, *The Birth of a New Physics.* Garden City, N.Y.: Anchor, 1960, ch. 6.
De Santillana, Georgio. *The Crime of Galileo.* Chicago: University of Chicago Press, 1955.
Drake, Stillman. *Galileo.* New York: Hill & Wang, 1980.
Dreher, J. L. E. *A History of Astronomy.* New York: Dover, chs. 14–16.
Koestler, Arthur. *The Sleep Walkers.* New York: Macmillan, 1959.
———. *The Watershed.* Garden City, N.Y.: Anchor, 1960.
Lodge, Oliver. *Pioneers of Science.* New York: Dover, 1960, lectures 2–3.
Pennekoek, A. *A History of Astronomy.* New York: Interscience, part 2: "Astronomy in Revolution," chs. 20–25.

8

Galilean Thought Experiments and Hypothesis Testing

When Galileo rolled balls down an inclined plane, a new light burst upon investigators of nature.

Immanuel Kant, *Critique of Pure Reason,* 1781

In the distance tower still higher peaks, which will yield to those who ascend them, still wider prospects.

J. J. Thomson, Nobel Prize speech, 1909

THE MYSTERY OF GRAVITY

An apple drops from a tree and falls to the ground. Why? We are so accustomed to seeing objects fall that we don't ordinarily think about it. We simply take the phenomenon for granted. If a child should ask why things fall, we might answer simply "gravity"—as if a label were an explanation. If pressed further, we might even say that gravity is an invisible force. But this is reasoning in a circle, like a dog chasing its own tail. We see an object fall and talk about a force; to prove the force, we point to the motion of the object. If this mysterious force does exist, what is it? How does it arise? From where? Can its existence be proved? Can something other than a force explain the fall of the apple?

THE WHY AND HOW

Thinking men speculated about falling objects even before Newton, Galileo, or Aristotle. But Aristotle was among the first to incorporate falling objects into a universal scheme of things. His ideas of gravity were part of his earth-centered scheme of the universe and supported it. There are the heavens and the earth.

115

In between is a matter-filled space. Not all substances fall when released; some rise. Burn wood, said Aristotle, and fire, air, and water will move upward; earthy ashes will remain behind. To him, earth, air, water, and fire were four basic elements that in different proportions composed all terrestrial objects. When an object burns, the elements within it are released to seek their own prescribed places in the order of things. The natural place of water is above the earth, air above water, and fire above air. Heavenly bodies are composed of a fifth, etherial element, or *quintessence,* not to be found below the moon.

Many terrestrial objects are made up mostly of the element earth. The natural place for earthy substances is the center of the universe, where the earth's center is located. When a rock hurtles down from a cliff top, it is seeking its natural center, much like a homing pigeon. Gravity is this downward natural "seeking" of earthly objects for their natural center; it is not a force from the outside. The sphericity of the earth is due to the drawing in of its earthy parts toward their natural place at the center of the universe.

Seas, rivers, and lakes, by their position, show that the natural place of water is just above the earth. Rain falls to its natural place above the earth. Underground springs are seeking their level when they spurt above the ground through a weak point in the earth that traps them. This anthropomorphic upward "seeking" was called *levity,* an upward gravity. Levity and gravity are understandable in terms not of weight but of elements animistically seeking natural places. The following hymn by Robert Seagrave (1742) is more than a metaphor; it reflects Aristotle's physics:

> Rivers to the ocean run
> Nor stay in all their course.
> Fire ascending, seeks the sun;
> Both speed them to their source.

This concept of five elements in natural places is the basis of a simple mechanics. Rest is the natural state of affairs for all objects in and around the earth. Everything has an appointed level in accordance with its composition. When an element or object is out of its place, a *natural motion* results when the substance is free to seek its natural level. No outside force is involved. In addition to this natural motion, there is a second type, a *violent motion,* caused by an outside force on an object. Throw a stone and you exert a force to give the stone a violent motion in a horizontal or even upward direction. This forced or violent motion is differentiated from the natural motion of the stone vertically downward. When in violent motions the stone can continue for a while in its horizontal or upward direction. The Aristotelian reason is that the air displaced in front of the moving stone rushes around it to act as a force behind it. As this process spends itself, the natural motion of gravity takes over.

Aristotle's physics is one of common sense. Observation shows that skimming pebbles or moving vehicles quickly come to a stop unless some push or pull is maintained. A constant force seems necessary for a constant velocity on the earth. Further, force and violent motion can be associated as *cause and effect.* That is, no force, no motion (just rest); no cause, no effect. On the other hand, force, a disturbing cause, results in violent motion, the effect.

Aristotle's ideas of motion are thus causal and nonquantitative. They attempt to explain *why* things rest or move rather than describe in any detail *how* things move. Terrestrial objects are at rest in natural places, in natural motion "seeking" their places, or being temporarily forced out of their places only to "seek" them again. The universe is an orderly, geocentric one. These ideas were effective enough to trouble Copernicans with such questions as how a stone thrown vertically upward can land on the same spot from which it was thrown if the earth moves under it. A moving earth should leave the air and stone behind. Another question was, how can the earth be the permanent center of gravity toward which heavy objects move and itself be moving around the sun? The earth is heavy and massive. It is at the center of the universe because that is the natural place toward which all heavy objects gravitate. A new science of mechanics was needed. And when Galileo was under house arrest during the last nine years of his life, it was to a new science of mechanics that he applied himself.

"When Galileo rolled balls down an inclined plane, a new light burst upon investigators of Nature," said Immanuel Kant. The "new light" recognized by the physicist and philosopher Kant was the dawn of a new mathematical approach to the universe now known as theoretical physics. Let us look further into the significance of Kant's words.

If a one hundred-pound boulder and a one-pound rock were dropped at the same instant from the top of a cliff, which would hit the ground first? About how much sooner? And why?

Aristotle's answer, in the *Meteorologica,* was that "bodies fall faster in proportion to their weight." That is, a one hundred-pound object should fall one hundred times faster than a one-pound object. Or, in Aristotelian words, "An iron ball of one hundred pounds falling from a height of one hundred cubits reaches the ground before a one-pound ball has fallen a single cubit." The commonsense reasoning was that to the extent the heavier object has extra weight, it is impelled to move faster. Snowflakes and feathers can be observed to settle to the ground at a fairly constant speed; raindrops, at a higher speed. Heavier objects than these fall faster because they are heavier.

Galileo questioned Aristotle's position and maintained that all objects, regardless of weight, would fall equally fast in a vacuum. Resistance of the air on wide surfaces holds back the feather and snowflake that in a vacuum would fall as fast as a coin or cannonball. Aristotle, claiming that "nature abhors a vacuum," did not think in terms of a vacuum. Galileo visualized a vacuum as an ideal, free-fall situation, without air resistance, and thereby equated the fall of objects. He backed up his reasoning in several ways. One method was to show through another simple thought experiment that two identical bricks would fall together whether separated or sides touching. Whether two bricks are touching to form one long brick or slightly separated to remain two, their downward velocity should remain the same; that is, objects of different weights should fall equally fast. More precisely, the velocity of a falling body should be independent of its weight.

That heavy objects would fall much faster than light ones seemed so self-evident to Aristotle and most of his followers that they felt it unnecessary to test the principle by actually dropping rocks. Records show, however, that John Philoponus, in the sixth century, did experiment with objects of different weights and found that they fell with almost equal speeds through the air. There is a story that Galileo dropped objects of different weights, sizes, and materials from the Leaning Tower of Pisa. While this story may be apocryphal, it symbolizes Galileo's experimental approach to nature. In any case, in his *Dialogues Concerning Two New Sciences* (1638) there is some evidence that he dropped objects from high places somewhere. Salviati, who represents Galileo in the *Dialogues,* does claim that "you find on making the experiment [with a one hundred-pound ball and a one-pound ball] that the larger outstrips the smaller by two finger breadths." Then a little later, another character, Sagredo exclaims, "But I, Simplicio, who have made the test, can assure you that a cannon ball weighing one or two hundred pounds, or even more, will not reach the ground by as much as a span ahead of a musket ball weighing only half a pound, provided both are dropped from a height of 200 cubits [about 300 feet]." Galileo explained that the slight difference in time between the two balls was due to air resistance. He claimed that with actual free fall in a vacuum, any two iron balls would hit the ground at the same time. Since then, this claim has been substantiated many times. Even a feather and a coin dropped together in a long vacuum tube arrive together at the bottom of the tube.

Actually, "free fall" as a concept involves an ideal situation in which objects can fall without any air resistance or friction* whatever. In actual life, even the best of vacuum tubes contains minute amounts of air or other gases. But Galileo's approach was to set up an ideal situation for the basic principle that all objects fall equally fast regardless of weight and then to make allowances for the air resistance and friction that exist in actual life. This "free fall" became an idealized conceptual model through which to understand small contact differences actually found in the speeds of falling objects. Beginning with Galileo, idealized thought experiments became established practice in probing nature.

---

*Friction is defined as a collection of forces that resist motion between surfaces in contact.

Thought experiments with falling objects in idealized situations gave Galileo his idea that weight does not influence the velocity of a freely falling object. But he did not stop there. In taking a new look at falling objects, Galileo interested himself in "mathematizing" and measuring their motion. The question was, How was this to be done? How could the fall or the motion of an object be reduced to numbers, so that more could be understood about gravity and about motion in general? Galileo's quantitative approach to nature here is quite well expressed by these words of William Thomson, Lord Kelvin (1824–1907):

> When you can *measure* what you are speaking about and express it in numbers, you know something about it; but when you can not measure it, when you can not express it in numbers, your knowledge is of a meagre and unsatisfactory kind; it may be the beginning of knowledge but you have scarcely, in your thoughts, advanced to the stage of science.[1] (Original italics).

Today we would say that Galileo was operational and mathematical. First of all, he believed that in understanding gravity, emphasis should be upon *how* things fall rather than *why* things fall. *How* is an appeal to observation; *why*, to speculation. Galileo believed that it is premature to speculate about the *why* before knowing more about the *how*. He therefore experimentally observed specific falling objects, but more than that, he systematically used mathematics as a tool, a languages, and a logic; he used it to measure, define, and describe what he saw.

MATHEMATICAL HYPOTHESIS TESTING: ROLLING BALLS DOWNHILL

A stone is dropped from a high cliff. If we were watching the stone, we would probably agree that its speed increases as it descends. Is the acceleration uniform or not? How could we find out? One way would be to ask ourselves *how* the stone would fall *if* it did fall with uniform acceleration, and then to determine whether it actually does fall that way. At least, that is the way Galileo approached the problem.

According to Galileo, in his *Dialogues Concerning Two New Sciences,* "a body is said to be uniformly accelerated when starting from rest, it acquires equal increments of velocity during equal time intervals." That is, if a car speedometer reads 5, 10, 15, and 20 mph after 1, 2, 3, and 4 seconds respectively, the car (starting from rest) has a uniform *increase* of speed or uniform acceleration of 5 mph.

Natural processes do not necessarily conform to our patterns of thought. It is one thing for Galileo to have spelled out an ideal of uniform acceleration, and quite another for objects to fall with uniform acceleration. For Galileo, the point was to put his idea to a test. As he expressed it: "But as to whether this acceleration is that which one meets in nature in the case of falling bodies . . . this would be the proper moment to introduce one of those experiments—and there are many of them—which [test] in several ways the conclusions reached."

Experiments are a way of asking questions of nature. Galileo tested gravity and set a research pattern that is typical of science even today. In this pattern, mathematical equations are used to predict events that can be checked experimentally. If the experimental results are as predicted, the equation is considered a good working hypothesis for further investigation. If not, the equation is abandoned, and a new hypothesis is sought. Let us consider Galileo's inclined-plane experiments to gain more insight into the nature of both gravity and modern science.

Galileo's hypothesis was that objects fall with a uniform acceleration g. Unfortunately, freely falling objects move so rapidly that even today it is difficult to measure their velocities directly. Falling objects just do not come with built-in speedometers. How could the velocity of an object at different points of its fall (or when it hits the ground) be determined? Instead of directly testing a freely falling object, Galileo sought an indirect test. He finally concluded mathematically that an object falls with constant increase in speed if the ratio of distance-of-fall (d) to time-of-fall (t) squared

retains the same value at all points of fall. Basically, all that is needed is a measuring rod and a timepiece.

In simplified, symbolic form, Galileo's reasoning was somewhat as follows: With a uniform acceleration a, distance is proportional to time-squared ($d \alpha t^2$) and forms a constant ratio

by considering that $d = v_{av} t$ and that $v_f = at$, where v_f is velocity at any instant of fall.

To determine whether the ratio d/t^2 is constant for freely falling objects, Galileo decided to "dilute" gravity and thereby increase the time of fall. He believed that gravity is a force causing a ball to roll down a hill as well as to fall freely in space. Instead of dropping objects through the air, he therefore decided to let a brass ball roll down an inclined plane at different angles.

At this point, we let Salviati in the the first dialogue describe for Galileo the actual experimental details and results:

We took a piece of wooden scantling, about 12 cubits long, half a cubit wide, and three finger breadths thick. In its top edge we cut a straight channel a little more than one finger in breadth; this groove was made smooth by lining it with parchment, polished as smooth as possible, to facilitate the rolling in it of a smooth and very round ball made of hardest bronze. Having placed the scantling in a sloping position by raising one end some one or two cubits above the other, we let the ball roll down the channel, noting, in a manner presently to be described, the time required for the descent. We repeated this experiment more than once in order to be sure of the time of descent and found that the deviation between two observations never exceeded one-tenth of a pulsebeat. Having performed this operation until assured of its reliability, we now let the ball roll down only one-quarter of the length of the channel; and having measured the time of its descent, we found it to be precisely one-half of the former. Next we tried other distances, comparing the time for the whole length with that of the half, or for three-fourths, or indeed for any fraction. In such experiments, repeated a full hundred times, we always found that the distances traversed were to each other as the squares of the times, and this was true for *any* inclination of the . . . channel along which we rolled the ball. (Italics added)

Fig. 8.1 illustrates the results described above by Galileo in letting a ball roll down an inclined plane. Notice that the.total distances from the starting position have the relationship of 1:4:9:16:25, or 1^2:2^2:3^2:4^2:5^2, at the end of the first, second, third, fourth, and fifth seconds. Thus

$$\frac{d}{t^2}$$

does retain a constant value here or $d \propto t^2$. This relationship between distance and time holds regardless of the angle between the plane and the horizontal. Galileo's actual results thus did match his predicted results and his mathematical hypothesis of a uniform acceleration of gravity was verified, at least tentatively.

Also again notice that Galileo in his description is idealizing his results. Here, as in free fall, he is reporting his results as if he had operated in a vacuum with a perfectly smooth ball on a perfectly smooth inclined plane.

Galileo's procedures in his famous inclined-plane experiments are summarized below.

1. Galileo posed the question of whether objects fall with uniform acceleration.

2. He sought an answer by *assuming the hypothesis* that they do fall that way.

3. He mathematically deduced from his definition of uniform acceleration a *constant ratio* between the distance of fall and the time squared(d/t^2) of objects affected by gravity.

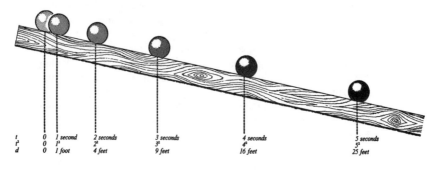

Figure 8.1

"Diluted" free fall. The distance *d* that a ball rolls down an inclined plane varies with the square of the time (t^2).

4. He tested his hypothesis of uniform acceleration through experiments in which balls were rolled down inclined planes in a way that permitted measurements of distance and time.

5. He repeated the experiments at different angles of the inclined plane.

6. He accepted his hypothesis that objects fall with uniform acceleration when he found that the measured ratio of distance and time squared was a constant as predicted for any angle of the inclined plane. Each angle of the plane with the ground had its own constant, however.

Galileo's inclined-plane experiments illustrate that science proceeds not only inductively from observation to ideas but also deductively from ideas to observations. Galileo first formed his hypothesis of a uniform acceleration *g* for gravity, and then he tested it by taking distance and time measurements of a ball rolling downhill. He was proceeding inductively from a general observation to an idea (mathematically expressed) and then deductively from the idea to further evidence—all in the same *ad hoc* experimentation. Both subjectivity and objectivity are at work here: subjectivity in the formation and formulation of the hypothesis; objectivity in testing the hypothesis.

INERTIA, MOVING MASTS, AND FALLING TARGETS

We previously discussed Galileo's use of free fall in a vacuum as an idealized conceptual model through which to understand small differences actually found in the speeds of all falling objects. Let us now conclude with several more of his significant thought experiments. When the ball in Fig. 8.2 rolls down the inclined plane on the left, it gains speed; indicated by the + sign. In rolling up the inclined plane on the right, the ball loses speed, as shown by the - sign. The ball has positive acceleration downward and negative acceleration upward. If the ball and the horizontal surface between the two planes are perfectly smooth, the acceleration sign would be neither a + nor a – but a zero on the horizontal surface. The speed would neither increase nor decrease but would remain constant on the horizontal.

This has great significance, said Galileo. If the frictionless horizontal plane were indefinitely extended, the ball would continue to roll on indefinitely without any force acting upon it. The inference was revolutionary. Under frictionless conditions, the ball would be in permanent natural motion. Motion in a straight line at constant speed is just as much a natural state of affairs as rest, reasoned Galileo. In fact, rest is just a special case of motion in which speed equals zero ($v = 0$). Objects ordinarily slow down because friction slows them down. The smoother the ice on a pond, the less the friction force, and the further a smooth pebble skids. Understand motion through the idealized frictionless situation, and then make allowances for the friction that does exist in actual situations. In such a way, idealized thought experiments can give greater understanding of nature. In any case, in this thought experiment of Galileo, we first see the modern concept of inertia, the concept that once any object is moving, it will continue on indefinitely at constant velocity until an unbalanced force acts upon it.[2] This was truly a revolutionary concept, one based upon an idealized thought experiment, and one that led to Newton's first law of motion, the law of intertia.

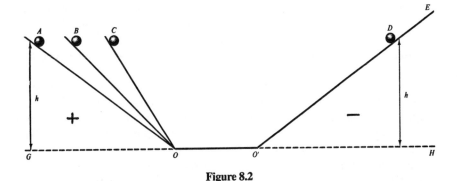

Figure 8.2

Galileo: "The speeds acquired by one and the same body moving down planes of different inclinations are equal when heights of those planes are equal." Whether starting from point *A, B,* or *C,* the ball rolls up to point *D* (if friction is negligible).

In light of the above, consider another of Galileo's thought experiments (Fig. 8.3). If a stone is dropped from the top of the mast of a ship at rest, no one is surprised if the stone lands at the base of the mast. But what if the ship is moving at 20 mph when the stone falls? Will the stone still land at the base of the mast, or will it land behind the base? Whether the ship is at rest or in motion, said Galileo, the stone will still land at the base of the mast. Before the stone was dropped, it was traveling horizontally at 20 mph along with the ship. Even though the stone is separated from the ship when it falls, the stone is still traveling at 20 mph by inertia, like the ball rolling along the horizontal plane. By inertia, the stone, even while falling, moves along horizontally with the ship until it hits the deck at the base of the mast.

Now, let the earth be the ship and the top of a cliff be the mast. If a stone is dropped from the top of the cliff, the stone, because of inertia, should hit the ground at the base of the cliff whether the earth under it is moving or not. The stone was on the moving earth (ship) before it was dropped and separated. Similarly, a stone thrown vertically upward eventually lands at the same place from which it was thrown, even if the earth under it moves. The stone was on the earth before being thrown and, by its inertia, persists in the same horizontal motion. Thus, even before actual measurements, with two qualitative thought experiments Galileo escaped from the restrictions of Aristotelian thought.

Now note in Fig. 8.3 that the path of the stone appears to be a straight (dashed) line on the ship, but a parabola from the shore. To a sailor who releases a stone from the top of the mast, the stone falls vertically. Only gravity seems to be acting. To an observer on shore, however, the path of the same stone is a curved, not a straight line. The motion of the ship is superimposed on the fall of the stone. Note the relativity of motion: A straight-line path for the sailor on the moving ship becomes a parabolic curve for the observer on shore. A single event from two frames of reference gives rise to two different observations. These two differing observations for the same event from two reference frames emphasize very well the interactive character between an observer and what is observed.

But whether seen traveling in a straight line or in a parabolic curve, the stone hits the deck for both observers. A vertical motion of free fall is there for both, whether seen as combined with horizontal motion or not. Galileo generalized this idea as a principle of *independence of velocities*. According to this principle, projectile motion can be resolved into two independent components, a vertical and a horizontal. The horizontal motion remains constant if air resistance is ignored, and the vertical acceleration of gravity occurs independently of the horizontal motion. In another example, an expert archer aims an arrow at an apple on a tree (Fig. 8.4). The line of sight is horizontal and the distance reasonably close. If the apple drops the instant the archer releases his

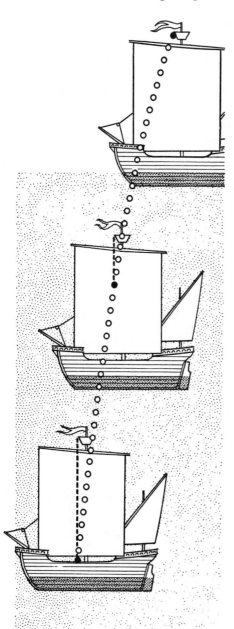

Figure 8.3

Law of independence of velocities. Whether the boat is at rest or in motion relative to the shore, a stone dropped from the mast top hits the deck at the base of the mast. By inertia the stone, even while falling, moves along horizontally with the ship until the stone hits the deck at the base of the mast. The path of the stone appears to be a straight line on the ship but a parabola from the shore.

arrow, the arrow hits the falling apple. The path of the arrow is a parabola. In approaching the tree the arrow falls vertically as fast as the apple and therefore hits it.

Centuries of experimental evidence supporting Galileo's hypothesis about projectiles have made his principle into a *law of independence of velocities*. In physical science, principles constantly supported by observational evidence tend to become laws. The law of independence of velocities, like the principle, states that projectile motion combines free-fall acceleration with uniform horizontal motion if air resistance is ignored. The vertical acceleration remains independent of any uniform horizontal motion. Air resistance slows down horizontal as well as vertical motion, but does not otherwise affect their independence. In Galileo's independence of horizontal and vertical velocities we find the germ of Newton's highly significant concept of vectors as well as a significant clue to understanding what keeps planets and satellites in orbit.

CONCLUSION

Imaginative ideas can be powerful tools. Galileo's thought experiments were based upon creative imagination, reasoning, and experience; they led to the revolutionary concepts of inertia and independence of velocities, which became pillars of Newton's monumental classical physics. Galileo developed mathematical principles of motion and tested gravity with them. In the process, there arose new concepts, a new body of knowledge, and a new method of inquiry. The new method of inquiry may be referred to as ad hoc mathematical hypothesis testing. Thought experiments and/or former experience lead to a tentative mathematical hypothesis, which is used to predict measurable

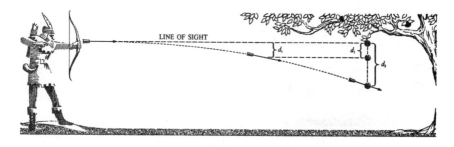

Figure 8.4

Testing the principle of the independence of velocities. If the target apple drops the instant the arrow is released, the arrow should hit the apple on its way down by the principle of independence of horizontal and vertical velocities.

events. Controlled experiments are then set up to check the predictions. Galileo's experimental appeal was to the mind, imagination, and senses. He used mathematics as a language and a logic that permitted measurement and evidence. Galileo, like Kepler, was delighted to find that again and again the universe showed itself to be mathematical. Gravity could be described and tested mathematically—with success. Galileo's idealized mathematical experimentalism characterizes science today. For example, we shall see later how Einstein used highly effective thought experiments in developing his theories of relativity.

Further, the process of hypothesis testing is clearly one of *interaction* between the scientist and nature. A Galileo observes a general increase in the speed of freely falling objects. This becomes a clue or signal *from nature.* He then wonders whether the increase in speed or acceleration is uniform. Through actual experimentation based upon thought experiments and idealized mathematical reasoning, he concludes that the acceleration of gravity g is uniform if *distance of fall*

$$\frac{d}{t^2} \text{ (time squared)}$$

= a constant at different distances of fall. By properly setting up inclined plane experiments, he addresses his *question* (of uniform acceleration) *to nature. Nature answers* in measurements of the rolling ball: If the measured results repeatedly match the predicted results of

$$\frac{d}{t^2}$$

= a constant, regardless of distance rolled (and at any downhill slope), Galileo receives nature's tentative, feedback answer of yes—or, at least, he interprets it as such. But interaction is still evident. Galileo's experiment serves as a two-way bridge of communication between himself (his ideas) and a natural event (a ball rolling downhill).

Notes

1. In Gerald Holton and Duane Roller, *Foundations of Modern Physical Science* (Reading, Mass: Addision-Wesley, 1958), p. 229.

2. We shall see later that thought experiments gave Einstein knowledgeable insights into the development of his relativity theories. It was truly a revolutionary concept based on an idealized thought experiment, and one that led to Newton's first law of motion, the law of inertia.

SUGGESTIONS FOR FURTHER READING

Aristotle. *Meteorologica*. Oxford University Press, 1923.
Beveridge, W. I. R. *Art of Scientific Investigation*. New York: Vintage, 1957.
Butterfield, Herbert. *Origins of Modern Science*. New York: Collier, 1962, ch. 5.
Cohen, I. Bernard. *The Birth of a New Physics*. Garden City, N.Y.: Anchor, 1960, ch. 5.
Galilei, Galileo. *Dialogues Concerning Two New Sciences,* trans. Henry Crew and Alfonso de Salvio. New York: Dover, 1952, "First, Third and Fourth Days."
Lemon, Harvey B. *From Galileo to the Nuclear Age*. Chicago: University of Chicago Press, 1949, chs. 1, 3–6.

9

Newton's Celestial Clockwork: Equations Predict Events

Great ideas emerge from the common cauldron of intellectual activity, and are rarely cooked up in private kettles from original recipes.
James R. Newman, *Scientific American*, June 1960

If I have seen farther . . . it is because I have stood on the shoulders of giants.
Isaac Newton, *Principia Mathematica*, 1687

Galileo and Kepler were the giants upon whose shoulders Newton stood. He started with Galileo's laws of falling bodies and Kepler's law of planetary motion and ended with a theory of gravitation that unlocked the universe. Newton's key was the concept of force. He mathematically related force to acceleration, to circular motion, to Kepler's laws, and to a universal gravitation. And when he was through, the cosmos seemed like a huge machine grinding away as relentlessly as fate.

WHAT HOLDS THE SOLAR SYSTEM TOGETHER?

It began in 1666, when the plague ravaging London was keeping Newton home on the farm in Lincolnshire. The story goes that an apple falling from a nearby tree caused him to raise questions about the moon. Does the earth's gravity extend to the moon? If so, why hadn't the moon ages ago fallen to the earth like a monstrous apple? Do the same laws of motion that apply to rolling balls or falling apples apply to celestial objects, too? If so, what holds the solar system together? These were basic questions that Newton addressed to the heavens. How he obtained answers in feedback from nature is what this chapter is about. In the process he established a classical scientific approach to nature. Further, he showed that the weight of objects, including our own weight, is to be understood not as an independent property but as an interac-

129

tion between the objects and the earth. As I see it, this opened the door to later concepts, theories, laws, and conceptual models as in Einstein's relativity, which had to be understood in terms of interaction. Newton's interaction concept of weight also opens the door, in my opinion, to the proposition that scientists themselves are parts of nature in interaction with their surroundings as they observe, explain, and predict events. But more of this later. Let us return to Newton's original questions.

Systematic answers to his questions comprise much of Newton's monumental work *Philosophiae naturalis principia mathematica (Mathematical Principles of Natural Philosophy)* (1687). The law of universal gravitation that emerged is one of the greatest syntheses in all science. Newton's thinking in the development of this law is classic in scientific research and deserves detailed consideration. The *Principia* is a formal presentation of Newton's ideas published twenty-one years after he developed them. We therefore cannot be sure of the sequence of his ideas as they actually occurred. Human insights and thought processes are complex. Often a flash of insight occurs first, and then logic and evidence are built up to support the insight. In any case, the sequence that follows relating Newton's momentous investigation of gravity is arranged for continuity and clarity of exposition. Just enough mathematical formulation is introduced to illustrate the significant role of mathematics as a language and logic in expanding interactions of scientists with nature.

A BIOGRAPHICAL SKETCH OF NEWTON

Isaac Newton was born in 1642, the year of Galileo's death, on a farm in Lincolnshire, England. His mother, widowed early, hoped that her only son would grow up to run their small farm. But the young Newton had little inclination for farming. Sundials and waterclocks were more appealing than horses and plows. Newton's uncle influenced his mother to send him to Cambridge University. There a new world of mathematics, astronomy, and optics opened up for him under the guidance of the eminent Lucasian Professor of Mathematics, Isaac Barrow.

When the plague decimating London closed Cambridge during 1665 and 1666, Newton retired to the comparative safety of the family farm. It was during the quiet of these eighteen months that the shy, reticent Newton, with ingenious insight, developed differential and integral calculus, formulated a new theory of color, established his famous laws of motion, and "mechanized" the universe in one of the great scientific breakthroughs of all time. His *Principia* dominated physics and astronomy for 250 years; and his *Opticks,* published in 1704, is still a model of scientific hypothesis testing.

Newton returned to Cambridge as a graduate fellow in 1667 and two years later became professor of mathematics when Barrow retired. At thirty, Newton became a fellow in the British Royal Society, the highest of scientific honors.

The remarkable creativity of Newton's twenties subsided in his thirties. *The Royal Society Transaction of 1672* contains a description by Newton of his new reflecting telescope. After that, the oversensitive Newton, disturbed by controversies with such eminent colleagues as Hooke, Huygens, and Leibnitz, seems to have lost enthusiasm for scientific investigation. It was only at the insistence of Newton's friend Halley that the *Principia* was finally published.

In any case, Newton was elected a member of Parliament by the university in 1688. Eight years later, he entered into permanent government service as Warden of the Mint; and in 1699 he became Master of the Mint, a position which he held for the rest of his life. Although Newton served as president of the Royal Society from 1703 to 1727, his own investigations during those years were primarily theological rather than scientific.* His early scientific achievements, meanwhile, had made him famous throughout Europe. When Isaac Newton died in 1727, he was buried in Westminster Abbey, internationally recognized and honored.

FORCE AND MOTION

Force is not a new concept. Men early identified force as push or pull that tends to move stationary objects. Then Aristotle emphasized that force is needed not only to move stationary objects but also to keep them moving. Without a sustaining force, a skimming stone slows to a stop. If planets and stars move ceaselessly, it is because celestial spheres carry them; and behind the celestial spheres is an Unmoved Mover.

That all motion would cease without intervening forces was disputed by Galileo. He claimed that objects at the earth's surface tend to move at constant velocity without forces. In Galileo's inclined-plane thought experiment (see Fig. 8.1), once the ball rolls down to the horizontal, it can be expected to roll along the horizontal indefinitely at constant speed if all friction is removed. Horizontally rolling balls or skimming stones ordinarily slow down only because friction forces resist motion. Remove friction, said Galileo, and objects once set in motion continue along endlessly. No force is necessary to keep them moving—only to stop them.

This revolutionary concept of inertia was the cornerstone of a new science

---

*At this time a scientific and industrial revolution was taking hold in England. Textile manufacturing was increasing, trade was expanding, cities were growing, and scientific communication was becoming international through scientific societies and their journals.

of mechanics that resolved the Ptolemaic-Copernican issue. In developing this new science, Newton continued from where Galileo left off, with three basic postulates that became laws of motion,

Newton's first law of motion, often called the *law of inertia,* states that *every body continues in its state of rest or of constant speed in a straight line unless acted upon by an unbalanced force.* In this statement, Newton is accepting Galileo's concept of inertia, but he is also extending it. By emphasizing *every body,* Newton is boldly applying the concept of inertia to the whole universe. Whether it be a ball, a stone, or a meteor, an object moves at constant speed in a straight line not only on earth but anywhere in space unless an unbalanced force acts upon it. Newton's first law of inertia is dynamic, universal, and revolutionary. Motion, not rest, is the natural state of affairs. Rest is a special case of inertial speed, $v = 0$. A parked car has an inertial speed of zero and tends to remain at that speed.

The first law involves motion without force: No force, no acceleration— just rest or constant speed in a straight line. Newton's second law, however, provides for motion with force. As Newton expressed it in his *Principia,* "Change of motion is proportional to the motive force impressed, and is made in the direction of the right [straight] line in which that force is impressed." Clearly, *force is associated not with constant speed but with changing speed or acceleration.* To Newton, force is a cause of acceleration. Wherever an object is accelerating, a force is acting. A force may be assumed to act on a falling object, even if an arm cannot be seen stretching up from the earth to pull the object downward. Consider one of today's rockets coasting at constant speed in a straight line far out in space. If one of its engines is ignited, the rocket will accelerate (or decelerate). The *applied force* of the engine results in acceleration of the rocket.

Newton's second law, commonly known as the *law of acceleration,* may also be expressed as follows: The acceleration of a body of given mass is proportional to the applied force and is in the direction of the force. Clearly, this law does more than associate an accelerated mass and a force. It relates them on a precise, quantitative basis. In the symbolic language of algebra, the law may be expressed as

$F \alpha\ ma$, where F is a *force* acting upon

$\qquad$ m, a given *mass,*

$\qquad$ a is the *acceleration*

$\qquad$ and α means *proportional to.*

To illustrate:

F	α	a
If a 10-lb force gives a mass m an acceleration of		5 ft/sec²,
then a 20-lb force will give the mass m an acceleration of		10 ft/sec²,
and a 30-lb force will give the mass m an acceleration of		15 ft/sec²,
or a 5-lb force will give the mass m an acceleration of		2.5 ft/sec².

or

F	α	m
If a 50-lb force gives a	500-lb piano an acceleration of 3.2 ft/sec²,	
then a 100-lb force should give a	1000-lb piano an acceleration of 3.2 ft/sec²,	
and a 150-lb force should give a	1500-lb piano an acceleration of 3.2 ft/sec²	
or a 25-lb force should give a	250-lb piano an acceleration of 3.2 ft/sec².	

In the second illustration given above, the accelerating force F increases with the mass m of the piano. As in the first illustration, the acceleration will increase with the accelerating force on the same piano mass.

Weight is a force of gravity that acts on a mass tending to accelerate it. Your weight is not a property of yourself, but an unseen external force that tends to pull you to the center of the earth. The earth's crust prevents you from accelerating downward as you would on water, snow, or quicksand.

As we shall see, Newton successfully extended his concept of accelerating force to the universe, and the equation $F = ma$ became a law. From Newton on, laws in the physical sciences increasingly took the form of equations. Mathematical physics at least was here to stay. And as Einstein expressed it:

> Physics really began with the invention of mass, force, and an inertial system. These concepts are all free systems. They led to the formulation of the mechanical points of view. . . . Science is compelled to invent ideas corresponding to the reality of our world. . . . Science is not just a collection of laws, a catalogue of unrelated facts. It is a creation of the human mind, with its freely invented ideas and concepts.[1]

The creation, I believe, flows from interaction between the human mind and its surroundings.

Forces always occur in pairs. This idea, uniquely Newton's, was expressed as follows in his third law, *the law of action and reaction*: "To every action there is always opposed an equal reaction; or the mutual actions of two bodies upon each other are always equal, and directed to contrary parts." Expressed briefly Newton's third law states: *Every force involves an opposite and equal reacting force.* In other words, wherever one object is found exerting a force upon another, the second object will be found exerting an equal and opposite force back upon the first object. Algebraically, $F_1 = F_2$. The forces of action and reaction do not act upon the same object but upon the two objects that are interacting. A few illustrations follow.

Step out of a rowboat onto a dock. The boat moves backward as you move forward. Your foot and the boat are interacting objects. The *backward force* of your foot *upon the boat* is matched by the *equal forward force* of the boat *on your foot.*

In walking, you press each foot against the ground to exert a *backward force upon the ground.* This is accompanied by an *equal forward force upon your foot* by the ground as you move forward. Now the ground and foot are the interacting objects exerting opposite and equal forces. If "terra firma" (the solid earth) is perfectly smooth ice, you are unable to exert a force on it, and it on you. You therefore slip and fall; you are unable to walk.

The hot gas leaving a jet plane in one direction is accompanied by the plane moving in the opposite direction—action and reaction.

When a cannonball zooms forward, the cannon lurches backward. If on wheels, the cannon rolls backward when fired. But the body of the cannon is so much more massive than that of the cannonball that it does not acquire the speed of the cannonball.

In summary, Newton's three laws of motion are among the most important laws in physics. The second law establishes mathematical relationships among force, mass, and acceleration ($F = ma$) in a universe of motion. But even more basic than this, the two laws together establish precise working definitions of terms that they relate: force, inertia, and mass. Acceleration had been mathematically defined before Newton. Inertia with Galileo and Newton became the tendency of an object to maintain a constant speed in a straight line or to remain at rest. Force became that which when applied to an object tends to make it change its speed or direction. If inertia means maintaining the status quo of uniform motion or rest, a force means changing the status quo. An object accelerating anywhere indicates a force. Newton's third law is significant not only for its assumption of that force existing in pairs, but also for its emphasis on objects interacting one upon another.

HYPOTHESIS OF AN INVISIBLE GRAVITATIONAL FORCE

The falling apple hits the ground. If the moon "falls," it falls perpetually—it never reaches the earth. In considering this difference, Newton recognized that before falling, an apple is at rest; the moon is not. As a clue, recall that Newton had the concept of inertia. If the same laws of nature hold for the heavens as for the earth, then by the force of inertia, the moon—or any planet, for that matter—should move out of its orbit at a tangent. Why doesn't it? Is it not reasonable to suppose that gravity extending as a force to the moon prevents the moon from

A) FORCE ON STONE

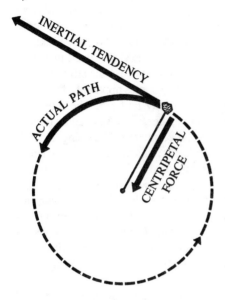

Figure 9.1

Inertia tends to send the stone off in a straight line, and the string to pull the stone toward the center. A circular path results. The inertial tendency of the stone results in its outward pull on the string

B) FORCE ON STRING

moving off at a tangent, much as a string acts on a whirling stone? (Fig. 9.1) That is, a stone moves in a horizontal circle when appropriately twisted by hand from a string. Inertia tends to send the stone off in a straight line, and the string tends to pull the stone toward the center by a force called centripetal (center-seeking). A circular path results. The inertial tendency of the stone shows itself in its outward (centrifugal) pull on the string felt at the hand. Release the string and the stone flies off at a tangent. The inertial tendency becomes apparent.

Through analogies a scientific hypothesis may be born. The moon replaces the stone; the earth, the hand; and gravity, the string. As in all analogies, however, there are differences between objects compared. The string, extending visibly between the stone and the hand, clearly pulls inward upon the stone. In the case of the moon, there is no observable material connection between it and the earth; after all, the distance between the two bodies is almost 0.25 million miles. How could a force be acting across such a distance? Newton's answer was simply that by analogy, the moon acts *as if* there were such a gravity force upon it across space—like that upon the apple when it leaves the tree! That is, by analogy, the moon's orbit could be the result of an invisible central gravitational pull of the earth that prevents the moon from moving off at an inertial tangent.

But Newton did not rest upon analogy alone. He mathematically proved that (1) any object moving at a constant speed in a straight line would travel in a circle *if* a central force continuously acted upon it, and (2) the object under this force would travel in accordance with Kepler's law of equal areas.

Newton's geometric reasoning in this is shown in Figures 9.2 to 9.5. For simplicity, we confine ourselves here to these geometric representations and the brief description underlying each.

In as simplified a form as possible the following algebraic proofs* is offered to illustrate Newton's further mathematical reasoning in the development of his hypothesis of a universal gravitation existing between all objects, including the earth and the moon:

1. Consider a planet (or the moon) to complete a circle C of radius in time T: $C = 2\pi r$

2. The speed v of the planet (or moon) is the circumference C (or $2\pi r$) divided by the period or time of a revolution T; that is, $v = \text{distance/time} = 2\pi r/T$.

3. Newton's derived expression for an assumed centripetal (or gravitational) force F of the sun acting on the planet is $F = mv^2/r$, where m is the planet's mass, v is the velocity, and r is the planet's orbital radius.

---

*In the next section, *mathematical proof* will be differentiated from *scientific evidence*.

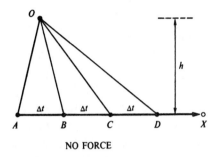

NO FORCE

Figure 9.2

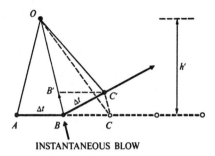

INSTANTANEOUS BLOW

Figure 9.3

No force and equal areas. An object moves with constant velocity along line *AX*. Another line from a central point *O* to the moving objects weeps over equal areas in equal times ($\triangle AOB = \triangle BOC = \ldots$

An *instantaneous* force and equal areas. A sharp blow toward *O* at point *B* sends the object toward *C'* instead of *C*. Equal areas still exist ($\triangle AOB = \triangle BOC'$) in equal units of time.

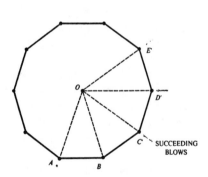

SUCCEEDING BLOWS

Figure 9.4

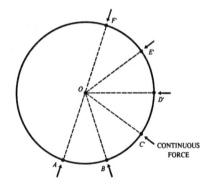

CONTINUOUS FORCE

Figure 9.5

A *repeated* force and equal areas. Further regular sharp blows toward *O* are made at points *D'*, *E'*, etc. The path of the object forms a regular polygon. Now $\triangle AOB = \triangle BOC' = \triangle C'OD'$, etc, in equal time intervals.

Continuous force and equal areas. A continuous force toward central point *O* instead of repeated blows gives the object a circular path.

Source: *Newton's Principia*; in James Perlman, *The Atom and the Universe*, pp. 174–75.

4. But by Kepler's harmonic law of planetary motion, $T^2 = kr^3$, where T is the period of revolution, k is a constant, and r is the orbital radius.

5. Algebraically combining $v = 2\pi r/T$ with the centripetal force equation $F = mv^2/r$ and with Kepler's third law $T^2 = kr^3$, the end result of $F = k'm/r^2$ is obtained, where F is the assumed gravitational force, m is the planet's mass, r is its radius of orbit, and k' is a constant.

The equation $F = k'm/r^2$ is Newton's famous inverse square relationship. This principle states, for example, that a gravitational force F from the sun upon a planet is inversely proportional to the square of the orbital radius r^2 of that planet. (The mass m of a planet is assumed to remain constant.) The greater the distance squared of the planet, the less the force of gravity.

By Newton's third law of action and reaction, a gravitational force from the sun to the earth means an opposite and equal gravitational force back to the sun from the earth. And the sun is farther than the moon from our planet. Thus, if the earth's gravity reaches the sun, it reaches the moon since the principle of inverse squares would also apply to the moon in its orbit; therefore, the earth's gravity theoretically extends not merely to the moon but to infinity. Place as large a number as you wish in the denominator of Newton's $F = k'm/r^2$; there will always be a value F for the earth's force of gravity. *Only if the moon were at an infinite distance would the earth's force of gravity become zero.* Similarly, a central gravitational force of the sun would reach the furthermost planet and beyond to infinity.

The inverse square relationship can be illustrated with your own weight, which is the earth's gravitational force upon your body. Through calculus, Newton also proved that a gravitational force may be considered to be operating from the center of mass of one object to that of another. The earth's gravity, accordingly, acts upon you from a center approximately four thousand miles below the earth's surface.

Let us say that your weight is 160 pounds. This means that a gravitational force of 160 pounds is acting upon you across a distance of 4,000 miles from the earth's center. In accordance with the inverse-square principle, your weight should decrease as you ascend above the earth (Fig. 9.6 and Table 9.1). For example, your weight at a distance of 8,000 miles above the earth's center is 40 pounds by Newton's inverse-square principle and at 16,000 miles, 10 pounds. Clearly, if Newton's inverse-square relationship holds, you would still have some earthly weight as far away as the moon. Only at an infinite distance from the earth would the earth's gravitational force cease acting on the moon or any other object. Thus, by Newton's "inverse squares" *your weight is not an absolute characteristic of yourself, but primarily a gravitational force of the earth upon you. It is an aspect of interaction between you and the earth in accordance with your mass.* This interaction is shown in proofs 6 and 7 below in Newton's law of universal gravitation. (See also Table 8.1.)

6. Further, by Newton's third law of motion: *Every action has an opposite and equal reaction;* the moon has an opposite and equal gravitational pull upon the earth symbolically expressed as

$F_e = -F_m$. Thus, $F_e = k'm/r^2$, the inverse square law,
becomes $F_g = GMm/r^2$ where M is the earth's mass,
$\qquad\qquad\qquad\qquad\quad m$ is the moon's mass,
$\qquad\qquad\qquad\qquad\quad r^2$ is the distance squared between them,
$\qquad\qquad\qquad$ and G is a universal gravitational constant.

7. But Newton did not limit gravitation to the earth and to the moon or to the sun and the planets. He extended it to all objects. The mutual gravitational force between the earth and the moon depends in part on their masses, that is, on the quantity of matter within them. Therefore, gravitation applies to all matter. Whether stars or grains of sand, all objects share a mutual gravitational attraction. A world gravity structures the universe and makes it one. Newton's crowning achievement, the *law of universal gravitation,* states simply that *every body in the universe attracts every other with a force that increases directly with each of their masses and decreases directly with the square of the distance between their centers.*

Shape, hardness, color, chemical composition—none of these matters in the universal gravitation of objects. We need to measure only masses and distance!

To summarize: In answer to the question "What holds the solar system together?" Newton extended Galileo's laws of falling bodies and his own laws of motion to moon and the planets, and he applied his (and Huygens's) concept of centripetal force to these bodies in orbit. Then with the aid of Kepler's laws, he hypothesized universal gravitational forces between the sun and planets and between planets and moons that tie them all together into one mechanical system dominated by the sun. The forces are mechanical; their effects are measurable and can be treated mathematically.

GRAVITATION TESTED: THE MOON'S
DAILY FALL TOWARD THE EARTH

Mathematical reasoning often leads to brilliant hypotheses about the world, but this reasoning does not constitute scientific evidence. For that, hypotheses must be tested. How were "inverse squares" to be tested? Newton's answer was simply this: Use the moon! Test the inverse-square relationship through

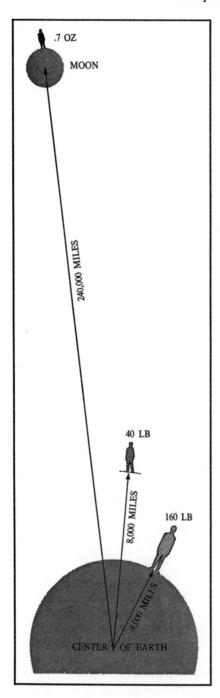

Figure 9.6

Weight decreases by an inverse square with distance from the earth's center (see Table 9.1).

Table 9.1
Inverse Square Relationship

Distance from Earth's Center

Miles		Inverse Squares	Your Weight Lb.	Oz.
4,000	(1 · 4,000)	$\dfrac{1}{1^2}=\dfrac{1}{1}$	1 · 160 lb = 160	
8,000	(2 · 4,000)	$\dfrac{1}{2^2}=\dfrac{1}{4}$	$\dfrac{1}{4}$ · 160 lb = 40	
12,000	(3 · 4,000)	$\dfrac{1}{3^2}=\dfrac{1}{9}$	$\dfrac{1}{9}$ · 160 lb = 17	12
16,000	(4 · 4,000)	$\dfrac{1}{4^2}=\dfrac{1}{16}$	$\dfrac{1}{16}$ · 160 lb = 10	
32,000	(8 · 4,000)	$\dfrac{1}{8^2}=\dfrac{1}{64}$	$\dfrac{1}{64}$ · 160 lb = 2	8
240,000	(60 · 4,000)	$\dfrac{1}{60^2}=\dfrac{1}{3,600}$	$\dfrac{1}{3,600}$· 160 lb = 0	0.7
∞		$\dfrac{1}{\infty^2}=0$	0 · 160 lb = 0	0.0

Clearly, if Newton's inverse square relationship holds, you would still have some earthly weight as far away as the moon. Only at an infinite distance from the earth would the earth's gravitational force cease acting upon you or the moon. The earth's gravity should thus extend to the moon by an inverse square relationship.

the moon's motion. First, *predict from inverse squares* how much the moon *should* fall daily; then *calculate from data* of the moon's motion how much the moon *does* fall daily. *If the predicted value closely approximates the observed value,* scientific evidence exists for an "inverse-squares" extension of gravity to the moon.

But how could the moon be falling to earth? The moon's path obviously is around the earth and not directly toward it. Consider Galileo's analysis of projectile motion (Fig. 8.4). By application of the principle of independence of velocities to the moon, the moon, by inertia, tends to move off at a tangent at any point in its orbit. This motion at a tangent is similar to the inertial horizontal velocity of the cannonball in projectile motion. And just as there is a vertical fall of a cannonball independent of its horizontal motion, so there is also a gravitational free fall of the moon toward the earth independent of its inertial motion. The moon's actual orbit is a resultant of its two independent velocities, just as the parabolic curve of the cannonball results from its two

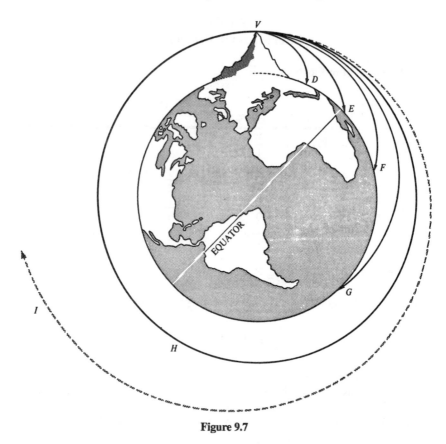

Figure 9.7

A replica of Newton's illustration in *Principia* explaining satellite motion through projectile motion.

velocities. The moon, therefore, is like a projectile circling around the earth but never landing because of the curvature of the earth. It is as if the earth bends back. This idea, highly fruitful with modern artificial satellites, was first recognized by Newton in the *Principia*. Through a famous illustration, he explains how a projectile fired with enough speed from a high mountaintop could become an artificial satellite or moon (Fig. 9.7).

Fig. 9.8 illustrates the *predicted* value of the moon's fall toward the earth. The moon by inertia moves 13.2° in one day from point M to point M'. But independently, the earth's gravity pulls the moon the distance d from S to M'. The distance d can be predicted by Galileo's law of falling bodies, $d = \frac{1}{2}gt^2$, where $g = .107$ in./sec^2 at the moon's distance from the earth. Calculation gives the distance d as 16 feet in one minute, or about 6,300 miles in one day.

Fig 9.9 describes the *observed* value of the moon's fall. The line *EM* is the average distance of 240,000 miles from the earth's center to that of the moon. *MR* is the straight-line path the moon would take if there were no gravitational force from the earth. The moon moves through the arc *MM'* during a period of one day. The angle *MEM'* is 13.2°, since the moon completes one revolution in 27.3 days and 360°/27.3 equals 13.2°. Since the scale of the drawing is 20,000 mi to 1 cm, the radii *EM* and *EM'* are approximately 12 cm. The line *SM'* represents the daily fall of the moon from a straight-line path. This line can be directly measured at about 30 mm, or 3 cm equivalent to about 6,000 miles.

The above theoretical prediction of approximately 6,300 miles for the moon's daily fall compares to about 6,000 miles for the observed fall. Similarly, when Newton found that his predicted values reasonably matched his observed values, he concluded that the force that pulls the apple to the earth is the same that keeps the moon in its orbit—a universal force of gravity.

INVISIBLE FORCES EVERYWHERE

Such was Newton's earliest evidence that a gravitational force of the sun holds the planets in orbit just as the earth's gravity holds the moon. Each planet would move off in a straight line by inertia if it were not for the sun and its central gravitational force. The moon, through "inverse squares," is therefore a connecting link between the gravity on the earth and the gravitational pull of the sun on the planets. Gravitation explains Kepler's laws; a follow-through of Galileo's new system of mechanics, it explains how the earth can be a planet moving around the sun. Gravitational forces between the sun and planets and between planets and moons tie them all together into one mechanical system dominated by the sun. Newton had given his answer to the question "What holds the solar system together?"

The earth's regular tides were also Newton's evidence that gravitation works both ways, that every action has an opposite and equal reaction: If the earth attracts the moon across space, then the moon, as well as the sun, must also attract the earth. Fig. 9.10 illustrates Newton's use of tides for evidence that gravitation works both ways.

If mutual gravitational forces exist among all objects, then why do not dishes on a table ordinarily gravitate toward each other like magnets? It is certainly reasonable to expect some evidence of gravitational attraction in everyday objects around us if gravity is in fact universal. Yet experimental evidence for the mutual attraction of ordinary objects was not found until about a century after the *Principia*. In 1798 Henry Cavendish (1731–1810), the eccentric genius after whom England's most renowned research laboratory is named, reported that he had measured gravitational forces among experimental objects

PREDICTION: d=½ gt²

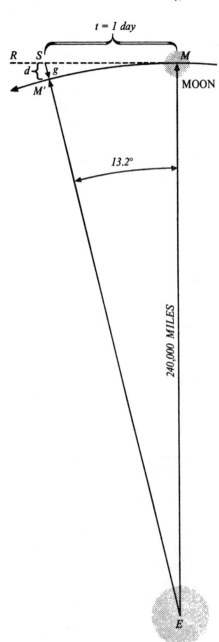

Figure 9.8

Predicted fall of the moon toward the earth: 6,300 miles per day. The moon, by inertia, moves 13.2° in one day from point *M* to point *S*. But independently, the earth's gravity *g* pulls the moon the distance *d* from *S* to *M'*. The distance *d* can be predicted by $d = \frac{1}{2}gt^2$ with $g = .107$ in./sec² at the moon's distance from the earth.

"OBSERVATION"

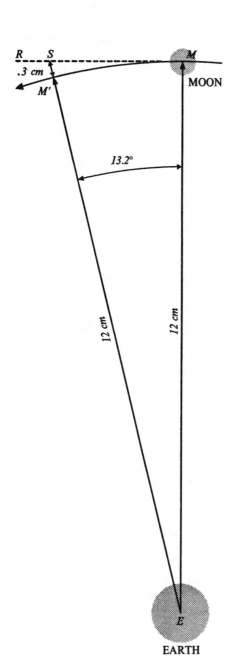

Figure 9.9

Observed fall of the moon: about 6,000 miles per day. By scale-drawing measurement, the observed fall of the moon *SM'* is about .3 cm, compared to 12 cm representing the 240,000-mile distance to the moon. Three-tenths of 1 cm represents about 6,000 miles.

of known mass. For his experiment, Cavendish had improved upon a torsion balance devised by a friend, Rev. John Mitchell. The torsion balance technique essentially was one of suspending from a very thin wire a long, narrow rod with small spheres, m_1 and m_2, at each end and then of bringing a pair of massive spheres, M_1 and M_2, toward the suspended spheres (Fig. 9.11).

As the large spheres came close to the small ones, the suspended spheres moved toward the larger spheres. The amount of gravitational attraction was detected by the measurable twist of the vertical suspension. Here was independent evidence for a universal gravitational attraction.

A few years later, the German scientist Philipp Johann Gustav von Jolly (1809–1884) reinforced Cavendish's evidence for universal gravitation. He used an equal arms balance in an ingeniously simple technique described in the caption beneath Fig. 9.12.

Newton's successful moon-fall predictions and the experiments of Cavendish and von Jolly solidly supported Newton's gravitational-force equation. It apparently could be applied to the universe. The hypothesis of gravitation, as mutual forces acting upon objects across space, had enough specific evidence to be further generalized into a theory. But perhaps it was when the equation led to the discovery of unknown planets that it reached the status of a law of the universe. The testing for universal gravitation began with the moon, proceeded earthward to ocean tides and man-made lead spheres, and then reached outward to space beyond the farthest known planets.

In 1610, Galileo reported discovering four moons of Jupiter through a telescope. But it wasn't until 1781 that telescopes, in the hands of William Herschel, were powerful enough to make possible the discovery of another planet. For years Sir William had been systematically reexamining the heavens with his prized 10-foot telescope and adding to his fame as a great English astronomer through his discoveries of new stars, comets, and nebulae. Then one night another new object of unusual appearance showed itself. When on many successive evenings this irregular speck of light changed its position among the stars but did not disappear, excitement mounted; here was not another comet but a new planet! Uranus had been silently orbiting the sun unknown to man until discovered by talent and chance.

Through a few observations and Kepler's laws, an expected elliptical orbit of eighty-four years was plotted for Uranus. All went well for about forty years; the observed positions of the new "wanderer" conformed to the expected orbit. Then things began to go wrong. More and more, Uranus departed from its prescribed path. Since due allowances had been made for gravitational pulls of other known planets, the explanation of the discrepancies did not lie there. Some astronomers suggested that perhaps Newton's law of gravitation was failing for large distances.

But independently, two young mathematicians became intrigued by

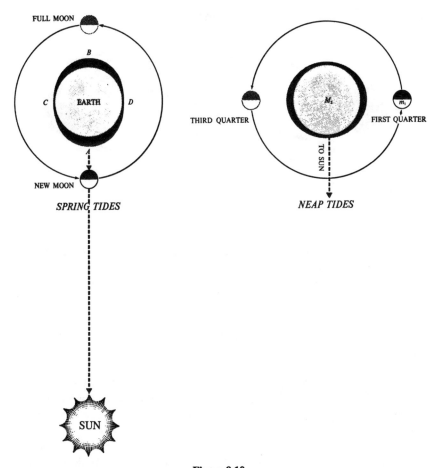

Figure 9.10

Spring and neap tides. Tides result from the gravitational pull of the moon and sun on the earth. Very high (spring) tides result at new and full moon when solar tides coincide with lunar tides. Much lowered (neap) tides result when solar tides are out of step with lunar.

another possibility: Perhaps another unknown planet beyond Uranus was pulling the latter out of orbit; perhaps Uranus was approaching an opposition with a new planet. If so, instead of being limited or discredited, Newton's law was being further confirmed. Each man—John C. Adams, a mathematics undergraduate at Cambridge, and J. J. Leverrier in France—made extensive calculations with Newton's law on the assumption that an unknown planet existed. In 1845, Adams, on the basis of his hypothesis and calculations,

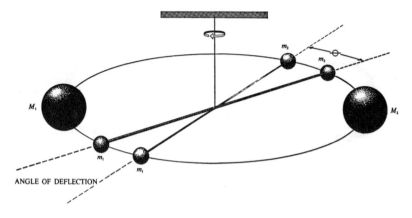

Figure 9.11

Cavendish's torsion balance. As the large spheres were brought close to the small ones, the suspended spheres moved toward the larger, the gravitational attraction being detected by the measurable twist of the vertical suspension.

alerted the Royal Observatory at Greenwich to train its telescope on a certain position in the sky. Unaware of Adams's efforts, Leverrier made the same request a few months later to the Berlin observatory. Adams's prediction was set aside at the Royal Observatory; the head astronomer at Berlin, however, acted immediately—and there the new planet was not far from the predicted position! On the basis of Newton's law of gravitation, the location of a new planet had been predicted before the planet had even been seen. Newton's gravitation was truly coming into its own as a law of the universe, and laws in the physical sciences increasingly were assuming the form of equations.

The new planet, Neptune, was the eighth. After a number of decades, this outer member of the solar system also began to wander from its expected positions. At the beginning of the twentieth century, the American astronomer Percival Lowell (1855–1916) invoked Newton's law to predict still another planet. The search lasted twenty-five years. Pluto was finally sighted in 1930 from Lowell's observatory; it was too small and too distant for earlier discernment. Who knows what other planet or planets may today be orbiting beyond Pluto?

NEWTON'S HIGHLY SIGNIFICANT PROBLEM-SOLVING PROCEDURE

If Kepler and Galileo initiated the use of algebraic hypotheses in testing hypotheses, Newton made explicit and established procedures that became classic in developing equations that became scientific laws.

What Holds the Solar System Together?

Newton's highly significant procedures for investigating universal gravitation are analytically summarized below: Newton

1. *assumed gravitational forces.* The earth's gravity was assumed to be a central force acting at a distance upon the moon by analogy with a stone on a string. A gravitational force from the sun was likewise assumed to act on each planet.

2. *geometrically deduced the possibility of central gravitational forces* from Kepler's law of equal areas.

3. *developed an equation to represent gravitational forces.* An "inverse square" equation for gravitational forces resulted from combining Kepler's third law and Newton's centripetal force equation. (Newton developed the centripetal force concept independently of Huygens.)

4. *algebraically synthesized a universal law of gravitation* from his "inverse square" equation and third law of motion.

5. *tested his universal law of gravitation* by

   a. *predicting the moon's fall* from Galileo's law of falling bodies, $s = \frac{1}{2}gt^2$, and Newton's "inverse square" law;

   b. *determining the actual "fall"* of the moon *from observed data;*

   c. *comparing results predicted from theory* in 5a *to observed results* in 5b;

   d. *accepting universal gravitation* when observed results closely matched predicted results;

   e. *finding further gravitational evidence* in tides.

   Note that points 1–4 in the description of Newton's scientific research above essentially involve *formulating a mathematical hypothesis* of universal gravitation whereas the fifth or last point emphasizes *testing that hypothesis* through prediction and observation as described in previous sections.

Newton's "Clockwork" Universe

Newton's laws of motion and gravitation explained the mechanics of the Copernican universe as Aristotle's principles had explained those of the Ptolemaic universe. Copernicus himself and even Kepler did not have physical explanations adequate for their systems. With Newton, however, "sun-cen-

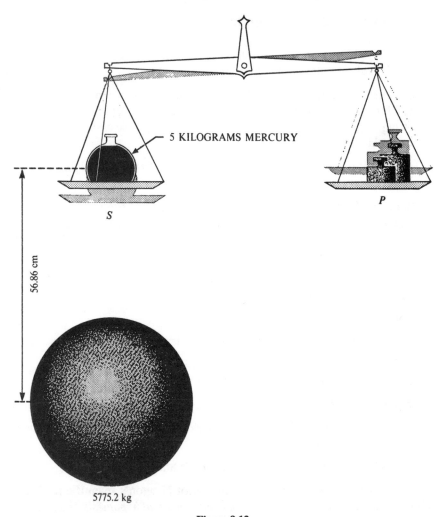

5 KILOGRAMS MERCURY

56.86 cm

S

P

5775.2 kg

Figure 9.12

Von Jolly's gravitational balance. A flask of mercury is thrown out of balance by a huge lead sphere placed underneath the pan *S*. The amount of additional weight placed in pan *P* to again balance the mercury equals the gravitational force of the sphere upon the mercury at the given distance.

tered" astronomers no longer needed to be defensive; they had acquired theoretical foundations. Copernicans now had forces at a distance to replace celestial spheres. Newton's system of mechanics, embracing new concepts of inertia, universal gravitation, and centripetal force, could effectively answer the thorny Ptolemaic questions emphasized in an earlier chapter and raised again for review at the end of this chapter. What is more, the new mechanical system's concepts and theories were backed by equations (as laws) and by experimental evidence, even if not by ancient authority.

Even more, Newtonian mechanics broke the deadlock of a simple observational relativism. In earlier chapters we saw that the same celestial observations could be used as evidence for either the Ptolemaic or the Copernican theory. For example, the sun is observed to move across the sky, daily drifting behind the stars. This can be explained by either the sun moving around the earth or the earth rotating and revolving around the sun. In either case, the result is the same. As long as there is relative motion between the earth and the sun, *to an observer on earth,* the sun appears to be moving around the earth. And yet, to make relativistic matters worse, *from the sun,* the earth would appear to be moving around the sun; *from the moon,* the earth would appear to be moving in a monthly cycle relative to the stars. All here seems to depend upon the observer's frame of reference.

But Newtonian mechanics provided an answer to the relativistic dilemma: actually, the smaller mass must move around the larger mass. The sun and the earth have equal and opposite gravitational forces upon each other to pull each other around a common center of mass—that is, the center of mass of the earth and the sun as a system. But the sun has 330,000 times the mass of the earth and therefore 1/330,000 the earth's acceleration. The resulting effect upon the sun is therefore negligible. Moreover, the common center of mass is quite close to the sun's center. The earth, therefore, moving around the common center of mass, moves around the sun. For the same reasons, the moon moves around the earth. Since the moon has only 1/81 of the earth's mass, the center of mass of these two as a system is near the earth's center. The moon, therefore, in moving around a common center of mass close to the earth's center, moves around the earth. If one accepts Newton's laws of motion and gravitation, one accepts the above explanations—and the Copernican system. The Royal Society of London gave its acceptance quite early.

Soon telescopes and the new physics not only displaced the earth but also the sun from the center of the universe. Galileo had found far more pinpoints of light in the Milky Way than men had ever dreamed of, let alone seen. Newton's friend Edmund Halley (1656–1742), for whom Halley's Comet is named, found the solar system to be moving relative to the stars. William Herschel in 1803 observed "double stars" revolving around each other. These "double star" revolutions were explainable by gravitation. Before long, the sun became merely one star among billions in a galaxy called the the Milky Way, which is also held together by gravitation. More than that, the sun was placed not at the center of the Milky Way, but about 30,000 light-years off-center. The Milky Way, with a diameter of about 100,000 light-years and a thickness of about 15,000 light-years, is lens-shaped, very much like its neighbor galaxy, Andromeda. The sun and its planets revolve around the galactic center at a speed of about 150 miles a second to complete one revolution in approximately 250 million years. But even our galaxy is not at the center of the universe; from recent observations, it is just one of billions of galaxies moving away from one another in an apparently expanding universe.

No wonder medieval thinkers feared for the Ptolemaic universe! But let us be careful that we do not appear medieval to men of the future. We, too, have an infinitely long way to go in knowledge, apparatus, and perspective.

EQUATIONS "EXPLAIN" AND PREDICT EVENTS

To explain and predict events in an uncertain world around them, men have always developed large conceptual schemes, whether in ancient mythology, Aristotle's concentric spheres, the Ptolemaic systems, Copernicus's sun-centered circles, or Kepler's elliptical orbits. Before Newton, conceptual systems or models of the world were geometric models. Even Copernicus had used geometric space relationships of deferents, epicycles, and eccentrics to make celestial predictions. Kepler and Galileo set beginnings for algebraic expression of laws, but with Newton, equations became an established form of hypotheses to be tested and of physical laws. Equations as mathematical models were found to describe and predict events more concisely and adequately than other models. Geometric models alone became insufficient.

Newton's law of gravitation is, of course, an equation. Gravitational forces at a distance may not exist, but Newton set up his equation on the assumption that they do. Forces are assumed to exist between any two masses by virtue of their being matter. If two masses and the distance between them can be observed and measured, the assumed force can be calculated although it is unseen ($F = G\,Mm/r^2$). From an animistic "homing instinct" of heavy objects for the center of the universe, Aristotle's gravity became, through Galileo, a local phenomenon of free fall on the moving planet earth. Now with Newton, gravity is a universal force represented by an equation which is a mathematical description of observed relationships between material objects everywhere. As an equation, gravity had a transition in status from hypothesis to theory and law. Gravity was a hypothesis when Newton assumed it to be a particular force acting from the earth upon the moon. As a hypothesis, it was a specific assumption to be tested with little or no previous evidence. Gravity was again specific, tentative, and limited enough to be a hypothesis when Newton used the concept to explain tides. In this case, gravity was a specific reactive force of the moon on the earth. And again, gravity was a hypothesis when Cavendish experimentally sought a "horizontal" gravitational attraction between metal spheres.

Gravity achieved the scope and relative certainty of a theory when it was tested successfully on an ad hoc hypothesis of forces acting at a distance in a variety of situations as above. The fall of the moon, ocean tides, and orbital fall of planets, when verified, were individual successes of the gravity hypothesis. Specific pairs of objects were involved in each case. The evidence could be generalized, even extrapolated, to consider gravity as a force existing not just between the earth and the moon, the moon and the earth, the sun and Mars, but between any two objects in the solar system or even in the universe. Accordingly, both a hypothesis and a theory are proposed explanations of events involving assumptions—for example, mechanical forces acting at a distance—but a theory generally involves more evidence, certainty, and scope than a hypothesis. Indeed, a theory generally includes several hypotheses.

Theories as explanations can be qualitative. Ocean tides, a falling apple, or a slight horizontal motion of Cavendish's small spheres toward larger ones can be qualitatively explained by the general theory of universal gravitational forces acting through distances: no numbers need to be used. Laws as equations have been defined by the logical positivists as *mathematical relationships among measurable characteristics repeatedly verified.* Newton's theory of universal gravitation became a law after the theory successfully took the mathematical form of an equation. Through mathematical logic and inverse squares, the theory was expressed in the quantitative relationships of an equation. As an equation, the theory still assumes universal forces acting at a distance, but it relates these unseen forces to measurable masses and distances. That is, the equation not only explains observed universal motion through gravitational forces but accurately predicts unknown future events through such assumed forces. The dramatic prediction and discovery of Neptune illustrates this point. Once the gravitational equation was solidly supported by evidence, it became a law.

Physical laws became universal with Newton. They must apply consistently and repeatedly everywhere—or else be qualified, modified, or replaced. Basic differences among hypotheses, theories, and laws, therefore, can be understood as differences in degree of certainty as well as in scope.

As we have seen, Kepler's laws had direct origins in Brahe's data. They were empirical laws. In comparison, Newton's law of gravitation developed in part from Kepler's laws and not directly from data. Kepler's laws mathematically described *how* planets revolve; Newton's law, in attempting to explain *why* they revolve, actually took a greater step into the *how.*

GRAVITATION AS A CONCEPTUAL MODEL

Newton's equations became components of a unified complex of even more scope than a theory or law: the conceptual model of a mechanical model universe. In this model, the universe appears as a huge machine grinding away as relentlessly as fate under Newtonian forces. It was the success of Newton's four laws in astronomy that led to their extension to all nature, the huge and the minute, in an all-encompassing mechanical view—i.e., that the universe can be understood primarily from concepts and laws of mechanics. This view was first expressed by Newton as follows:

> I wish I could derive all [phenomena] of nature by some kind of reasoning from mechanical principles: for I have many reasons to suspect that they [phenomena] all depend upon certain forces by which the particles of bodies are either mutually attracted . . . or are repelled and recede from each other.

Or, again, from the preface of the *Principia*:

> The whole burden of philosophy [of nature] seems to consist in this—from the phenomena of motions to investigate the forces of nature and then from these forces to demonstrate the other phenomena.

The universe became as deterministic as fate in the following famous passage written in 1802 by the French mathematician Pierre-Simone de Laplace:

> An intelligence knowing, at a given instant of time, all forces acting in nature, as well as the momentary positions of all things of which the universe consists, would be able to comprehend the motions of the largest bodies of the world and those of the smallest atoms in a simple formula, provided it were sufficiently powerful to subject all data to analysis; to it, nothing would be uncertain, both future and past would be present before its eyes.

In a mechanical view, all matter—and eventually energy, too—becomes composed of moving particles in interaction under universal forces. Everything is reduced to mathematical relationships among forces, matter, and motion in absolute space and time. And the forces derive from within matter itself. Even space is like a huge container in which all objects mechanically move about. The relentless determinism of a mechanistic universe is well demonstrated in a planetarium apparatus that projects with precision the appearance of the skies 2,000 years backward or 3,000 years forward. Know all causes, and effects are guaranteed.

A CROWNING SYNTHESIS

A great synthesizer takes the various pieces of a puzzle and fits them with penetrating insight into a common pattern—but many or most of the pieces have been shaped by other men. Newton was well aware of having related the work of Kepler, Galileo, Huygens, Descartes, and others; this was reflected in his reference to "standing on the shoulders of giants." But the pattern he came up with is one of the crowning intellectual achievements of all time. It dominated physics until the turn of twentieth century—about 250 years—and still serves well an appropriate realm in modern physics. Its mechanistic worldview, one of the great challenges in the history of ideas and technology, has changed the face of the earth.

Newton's synthesis, like all conceptual systems, represents, among other things, a search for order, unity, and simplicity. How else can a big idea unify a complex of observations, concepts, theories, and laws? Newton's synthesis also represents a search for natural causes, or at least a further step in the search. Newton's mechanical universe replaced celestial spheres with universal forces arising from within all objects themselves, whether in the heavens or on the earth. Forces acting at a distance displaced the animal-like "seeking" of objects for natural places.

Unseen forces may be a projection of man's mathematical mind upon the universe; even Newton stated that things act "as if" the forces are there. But certainly these forces were more impersonal than objects "seeking" places. In any case, forces were natural causes; and observations, effects. Unseen, impersonal forces explained elliptical orbits.

And the forces did so with equal signs. Newton's mathematical mind and faith was projecting equal signs upon the universe through gravitational forces—and the universe was apparently giving some support. A master synthesizer was *interacting* with his surroundings on a grand scale!

Not only was Newton projecting his grand model of gravitation upon the universe, but the universe was soon responding with favorable feedback by way of tides, observed "fall" of the moon, Cavendish's torsion balance, Jolly's gravitational balance, and sightings of new planets—Uranus, Neptune, and Pluto—predicted before being seen.

To celestial observations as old as man, Newton's gravitational forces linked concepts of inertia, principles of acceleration, laws of falling bodies and of projectile motion, laws of planetary motion, and a concept of vectors, as well as centripetal and centrifugal force concepts. Accompanying these were a host of assumptions, definitions, hypotheses, concepts, theories, principles, experiments, applications, and implications. Men of many nationalities through the centuries were represented, as they are in all science and culture.

In search of unity, Newton's mind and imagination related falling objects and celestial motions in a new, all-embracing order. To see likeness in unlikeness, to recognize similarity in the fall of an apple and the motion of the moon, is truly a unique quality of the probing human mind and imagination. As Alfred North Whitehead so aptly expressed it:

> The progress of Science consists in observing interconnections and in show-ing with a patient ingenuity that the events of this ever-shifting world are but examples of a few general relations, called laws. To see what is general in what is particular, and what is permanent in what is transitory, is the aim of scientific thought.[2]

And in the process—to use Rudolf Carnap's eloquent words—"The symbols and equations of the physicist bear the same relation to the actual world of phenom-ena as the written notes of a melody do to the audible tones of the song itself."

Once again, we see science intrinsically as a creative enterprise in human interaction with the surroundings—this time well symbolized by Newton's law of universal gravitation.

NOTES

1. Albert Einstein and Leopold Infeld, *The Evolution of Physics* (New York: Simon and Schuster, 1952).
2. Alfred North Whitehead, *Science and the Modern World* (New York: Mentor, 1949).

SUGGESTIONS FOR FURTHER READING

Bernal, J. D. *The Scientific and Industrial Revolutions*. Vol. 2. Cambridge, Mass.: MIT Press 1977, part 2: "The Birth of Modern Science."
Brownowski, Jacob. *The Common Sense of Science*. Cambridge, Mass.: Harvard University Press, 1977, chs. 1–4.
Butterfield, Herbert. *Origins of Modern Science*. New York: Collier paperback, 1962, chs. 5, 8, and 10.
Cohen, I. Bernard. *Birth of a New Physics*. Garden City, N.Y.: Anchor, 1960, ch. 7.
———. *Isaac Newton's* Principia. Cambridge, Mass.: Harvard University Press, 1971.
Conant, James. *Science and Common Sense,* New Haven: Yale University Press, 1962, chs. 1–3.
Feigel, Herbert, T., and Wilfred Sellers. *Readings in Philosophical Analysis*. New York: Apple-ton-Centurys Crofts, 1949.
More, Trenchard. *Isaac Newton—A Biography*. New York: Dover, 1962.
Pannehoeck, A. *A History of Astronomy*. New York: Interscience, 1961, ch. 26: "Newton."
Perlman, James S. *The Atom and the Universe*. Belmont, Calif.: Wadsworth Publishing Co., Inc., 1970.
Westfall, Richard S. *Never at Rest—Biography of Isaac Newton*. Cambridge, Mass.: Harvard University Press, 1987.

Part Three

The Human Dimension in Science: Problems of Scientific Objectivity

10

Expanding Partial Pictures:
Motifs, Models, Metaphors, and Hypothesis Testing

> The present laws of physics are at least incomplete without a translation into terms of mental phenomena. More likely, they are inaccurate. . . . As we consider situations in which consciousness is more and more relevant, the necessity for modifications of the regularities obtained for inanimate objects will be more and more apparent.
>
> Eugene Wigner, *Scientific American,* 1981

This chapter focuses further upon the human dimensions in science, that is, upon subjective, personal, and projective as well as objective aspects.*

We have seen that modern theoretical science had its origins with stargazers of old. The heavens suggested universal order. Celestial objects orbited regularity. And long before Ptolemy, ancient astronomers projected celestial spheres upon the heavens to house heavenly objects and their motions. Such spheres were symbolic: geometry was shaping astronomy. By projecting spheres first as conceptual frames of reference of moving points of light, and then as majestic rotating star-bearing structures, astronomers were not detached observers. They were interacting with and intervening in nature. They were "calling the world to order." Earth-centered world systems resulted: the models unified, explained, and predicted myriad observations successfully.

Previous chapters also traced transformations of grand conceptual models from Eudoxus to Newton in considerable detail. The perspective and the detail enabled many insights into the investigative character of science itself: a search for order, guiding principles, frames of reference, conceptual model building, hypothesis testing, observer-nature interaction, and perpetual reorganization of knowledge. The sun-centered model of Copernicus eclipsed the

---

*This chapter has to do primarily with *encapsulation,* i.e., conditioning by nationality, culture, religion, etc. This is explored fully in chapter 13.

earth-centered system of Ptolemy, only to be synthesized in the twentieth century into a larger Einsteinian framework. But this did not happen before Copernicus's circular planetary orbits gave way to Kepler's elliptical orbits and Copernicus's sun lost its favored position in the center of the universe. Each conceptual model in its time was highly innovative and productive in unifying and predicting myriad, fragmentary, and diverse observations. Each was based on reason and observation. In addition, each was human, even subjective, in the motifs, guiding principles, imagination, intuitions, assumptions, projections, and encapsulations they contained. The whole process—the use, development, modification, and decline of a conceptual model and its eclipse by a contender—appears as a perpetual reorganization of knowledge, perhaps a permanent scientific revolution.

Relativity, we shall see, is a current conceptual model that has absorbed the Ptolemic and Copernican-Newtonian models into itself. These earlier models are three-dimensional partial pictures of Einstein's larger four-dimensional model. The earth is an organizing center from which thinking, observing, imaginative men worked their way from immediate, personal, partial images of the universe to a more common space-time model. And men can work outward only from where they are. Thus, through big, unifying ideas, scientific man makes his way in a search for order in a complex, changing world. His big ideas run into contradictions, face more promising ideas, and are eclipsed. But each big idea has its own expansive effect. The broadening effect of Einstein's model in specific concepts of space, time, matter, energy, gravity, or geodesics is a case in point with its frame-of-reference approach to the universe. It is the best fit so far. In turn, it may become a facet in a still larger framework; but meanwhile, like previous models, it helps man creatively project himself outward and interactively gain new insights in the process.

In short, due to the limitations, comparative ignorance, and encapsulations (i.e., limitations in the knowledge process, to be discussed in chapter 13) of men; the vastness of the universe; and the interactive nature of the knowledge process, apparently separate scientific revolutions—as the Copernican, Newtonian, Darwinian, relativity, quantum, and cybernetic revolutions—are but critical phases or high point syntheses of a single, expanding developmental process. In the process, successive struggles of contending ideas and perpetual reorganization of knowledge spell out an ongoing revolution involving men as investigators in interaction with their surroundings.

Further, man's conceptual models of nature are to be differentiated from nature itself. Men approach nature with their models and, by an interactive process, often obtain in time some measure of correspondence between their ideas and the outside world. Testing a hypothesis within a model, for example, involves reasoning and sensory evidence. Mathematical predictions must match independently observed events. The orbital position of Venus at a given time as predicted by Kepler's hypothesis of equal areas (see chapter 7) must coincide with Venus's position as actually observed. Kepler's hypothesis became a law only with much successful prediction. Thus a reasonable correspondence between an idea and events was achieved.

Yet, motifs, models, and mathematical hypothesis testing in science suggest that there is much more involved than reasoning and observational evidence even in the fabric of the hard-core physical sciences. While testing a hypothesis clearly involves predictive reasoning and observation, *forming* a hypothesis is another matter entirely. It is often just as much a mixture of guiding faith, aesthetics, intuitive leaps, basic assumptions, and personal determination. This was as true of Einstein as of the neo-Pythagorean mystic Kepler. The intuitive and imaginative, the human and the accidental (i.e., chance happenings) enter into the process of hypothesis formation and determine what is conceptually projected and tested. And what is projected for testing can determine the nature of both the experimental apparatus and the results. We see this in the wave-particle controversy over the nature of light. When the English physicist Thomas Young (1773–1829) used diffraction gratings to test Huygen's wave theory of light, the wave theory was upheld. When photoelectric cells were used to test Einstein's photon theory of light, the opposing particle theory was vindicated. More on this later.

In previous chapters, we have seen that working hypotheses, theories, and laws all contain assumptions and vary from each other in the extent of evidence or in scope. The conceptual model is a unified composite of all of these plus many more component assumptions, concepts, data, relationships, reasoning, and so on. Kepler's laws led to a revised Copernican world system. And behind this system is a guiding faith in an overall simple, mathematical design of the universe. The Newtonian model of the universe as a huge machine grinding away inexorably as fate, includes assumptions of inertia and gravitational forces acting across space, mathematical concepts of point masses and vectors, theories of free fall and centripetal force, Kepler's laws of planetary motion, and Newton's own inverse-square law. A host of data involving the moon, tides, planetary motions, and terrestial motion are also there. Einstein's assumption of relative velocities of all objects and an absolute speed of light; his relativistic concepts of space, time, and mass; his theories that mass and energy are equivalent and that light bends in a gravitational field; his $E = mc^2$ law and his general law of motion—all are shaping components of his four-dimensional, space-time model of the universe as are his supporting data.[1]

But behind conceptual models and their various components are guiding faiths, intuitions, and principles with associated basic postulates. Often aesthetic, intuitive, and subjective in character, these guidelines lie outside the realm of just "pure reasoning" with data. What to others have been guiding principles, primary assumptions, faiths, motifs, or insights in science, Gerald Holton, in his *Thematic Origins of Scientific Thought,* refers to as "themes or themata, a third dimension to observational experience and logical construction." For their importance in reexamining problems of scientific objectivity, let us look more closely at some of these guiding principles or motifs behind conceptual models and all science.

A first basic assumption is that a *natural order of some sort exists in the universe;* otherwise there would be outright chaos. Without a basis for relationships among objects and events, no science could exist. By assuming a natural order, even if it be a chance order, science can proceed to search for this order, perhaps find some measure of it and in part even create it. As we have seen, early astronomers assumed order based on observations of periodic motions of celestial objects and then formulated their own designs for successful explanation and prediction. Even today, thanks primarily to a quantum revolution in physics, laws of nature are expressed in terms of chance rather than certainty. Statistical laws are a form of order expressing mathematically probable relationships and probability of events occurring. Now, if the universe is actually one of chance based on random motions of molecules, is the statistical order actually in the universe or is this a matter of scientific man calling the world to order through statistical, that is, mathematical, models of his own design?

But that brings us to a second basic guideline or guiding principle in the scientific approach to the universe: that the order in the universe is *mathematical.* Whether they express it in geometric, algebraic, or statistical terms, physical scientists and cosmologists since the Pythagoreans have shown a basic confidence in a mathematical universe as shown by their conceptual models.

Recall that to the Pythagoreans, numbers were the elemental realities relating all things. Numbers were identified with geometric forms (space and matter) in the following way: The number 1 was a point or unit indicating position. Two points determined a line; three points, a surface; four points not on the same surface, a solid. Numbers were then related to an elemental chemistry by way of the geometric forms: The number 4, representing solids, applied to Empedocles' four basic elements: fire, air, water, and earth. These elements were explainable, respectively, as four different regular geometric solids: the tetrahedron, octahedron, icosahedron, and cube. The fifth geometric solid, the dodecahedron, later provided for a fifth aetherial element. The variety of things in nature was understandable in the relationships and transformability of numbers and geometric forms. After all, wasn't nature full of geometric forms as in snowflakes, rock crystals, metallic structure, wood grains, and the like? Today's chemistry has its equations, but, of course, algebraic equal signs were not systematically projected onto the universe of moving objects until Kepler, Galileo, and Newton in the seventeenth century, or onto a conservation law of matter until Antoine-Laurent Lavoisier in the eighteenth century. They were not applied to a conservation law of energy until Joule, J. Meyer, and Hermann von Helmholtz in the nineteenth century, or to a combined matter-energy conservation law until Einstein in the twentieth. Pythagorean numbers were also related to physics and chemistry by way of the "music of the spheres." Pythagoras, for example, recognized a relationship between the length of a vibrating string and the pitch of its sound. He also recognized a pleasing harmony of tones when a string vibrated in an exact whole number of segments: 1, 2, 3, etc. Aesthetic feeling and scientific quantitative analysis were thus related. Like vibrating strings, celestial spheres had regular frequencies of motion and a harmony of their own. A "music of the spheres," therefore, truly existed. It was Kepler's great delight to detect this music in his harmonic law that relates the regular motions of all planets through a common ratio

$$\left(\frac{T^2}{\overline{D}^3} \right)$$

of period-of-revolution squared (T^2) to average-distance-to-sun cubed ($\overline{D}^3$).

Kepler, like the ancient Pythagoreans, was confident that geometry and numbers were the essence of the universe. He was not successful, despite his zeal, in connecting the five regular solids of geometry with the then six known planets. The solids eventually failed as a model to adequately explain the size, number, and arrangement of planetary orbits. But Kepler did come up with the ellipse as a model for orbits with the law of equal areas and with the harmonic law, all mathematical relationships which he tested.

Galileo's and Newton's successes in mathematical hypothesis testing have been fully analyzed in chapters 8 and 9. Einstein's four-dimensional space-time continuum may also be seen as a supreme Pythagorean model based on non-Euclidean geometry and other modern systems of mathematics.

Man's mathematics is still beautiful today when it applies to various realms of nature. It may turn out to be a huge metaphor. Newton was sophisticated enough to indicate that objects moved universally *as if* there were gravitational forces acting across space. If such forces existed, then the equation

$$F = \frac{GMm}{d^2}$$

expressed the magnitude of any such force acting across space between the centers of two masses. Einstein's concept of gravitation is in terms of a continuous four-dimensional space-time field, not of individual forces acting across three-dimensional space. If Einstein is correct, then Newton's gravitational force equation is a metaphor that works only under certain conditions. Will Einstein's space-time continuum also turn out to be another brilliant metaphor? Thus metaphors or other aesthetic guiding faiths have existed in the past and continue to exist as principles or motifs behind conceptual model building in science.

If it is a metaphor, mathematics has been a "desirable" one. Mathematical symbols and processes provide a language and logic that are emotionally uncharged—to satisfy one basic aspect of objectivity. Yet to be objective involves being detached in more than an emotional sense. A metaphor is not detached if it means a human conceptual projection upon the universe. In part, man may be "meeting himself" as he orders nature (Heisenberg). Problems of scientific objectivity will be reexamined in further chapters of this book.

A third basic guiding principle that modern scientists have retained from ancient natural philosophy is that of *simplicity*. Also called the principle of parsimony, it holds that other things being equal, the simplest explanation is the best. The Greeks, beginning with the Ionians, strained for a few unifying principles to explain all nature. Copernicus argued that his system would reduce the number and size of epicycles needed in planetary orbits. Kepler's ellipses superbly simplified planetary orbits, epicycles and all. The first of Newton's four basic "Rules of Reasoning," guiding principles to his system, emphasized the simplicity of nature and the necessity of explaining what is seen with no more hypotheses (assumptions) than necessary. Einstein's theories of relativity magnificently illustrated the guiding principle that a scientific system of ideas should rest on the fewest possible assumptions or postulates. Simplicity remains a basic criterion for evaluating a conceptual system. The guideline of simplicity has earmarks of the aesthetic and imaginative as well as of the convenient and the reasonable. Said Newton in an unpublished manuscript, "Truth is ever to be found in simplicity, and not in the multiplicity and confusion of things. . . . It is the perfection of God's work that they [i.e., the things of this world] are done with the greatest simplicity."

The universe, however, may not be simple. In the development of the concept of chemical elements, observation leads from the Ionian one-element concept to Empedocles' four-element theory of air, earth, water, and fire to the present periodic table of over one hundred elements. The process has gone from the simple to the complex. In the past generation, hundreds of nuclear particles have been discovered in the smashing of "elemental" atoms. A struggle continues to devise a suitable conceptual scheme that would bring order to a seemingly chaotic mass of data. Research in the realm of the submicroscopic has shown the universe to be far more complex than the *simplicity originally imagined.* That is the point; simplicity as a guiding principle may be a product of human imagination as well as of reason that may be reaching its limits of applicability. With man's comparative ignorance in a vast, dynamic, and possibly complex universe, a guiding principle of simplicity was a necessary start to knowledge—but only a start. Even if the principle should be reaching its limits of success, it has nonetheless served well in the development and continual reorganization of scientific ideas thus far.

A fourth guiding principle in science is *causality*. Every effect that we observe supposedly has its cause, obvious or hidden. Seek and you may find it. In Aristotle's ordered universe, a ripe apple falls to the ground when it, as a solid, becomes free to seek its natural place, the earth. Water seeks its level above the ground. The cause of any natural motion of an object is its "homing" tendency when, having been out of its natural position in the set scheme of things, it is free to return. Or again, the air displaced in front of a hurled stone rushes around the stone to act as a causal force behind it. Even the outermost sphere of stars had a Prime Mover as a Primary Cause behind its rotation. For Galileo, an apple fell because an external force of gravity acted on it from the earth. For Newton, an object accelerating anywhere in the universe had an unseen gravitational force acting upon it from somewhere across space.

God has been a causal motif to scientists wherever the limits of knowledge happen to be at any time. Aristotle's Prime Mover was behind rotating spheres. Newton's corpuscles (i.e., atoms) were indestructible because "God made them that way." Einstein's God did "not play dice with the universe." A strictly causal universe was a guiding principle.

Einstein directed the above words to those modern physicists whose guiding principle is a statistical universe, one of chance rather than of design. Yet even such physicists recognize patterns of events structured by chance. The probability of the right combination of physical circumstances to have occurred by chance in the early history of the earth for the emergence of life may have been exceedingly small, but once occurring, it set in motion a change of events that Darwin called evolution or change by natural selection. Thus, explanatory causes vary with the state of knowledge, individual inclinations, and conceptual model development at any given period.

The concept of *change* itself constitutes a fifth guiding scientific principle or motif dating back at least to ancient Ionian natural philosophers. Anaximander recognized transformtions of matter through a dialectical separation and "strife" of such opposites as moist and dry or hot and cold in a primordial transforming substance within all matter. Heraclitus's primordial fire of everlasting driving character was a primary element within substances transforming various substances, such as wood, into a host of other elements. The alchemists (unsuccessfully) and modern chemists (successfully) have been guided by a basic confidence in the transmutation of elements. Differences lay in the conceptual models and techniques that chemists and nuclear physicists were able to design.

Likewise, scientists had a basic confidence or guiding faith in the evolutionary change behind Darwin's theory of natural selection. And when eventually Darwin's natural selection alone could not adequately explain the origin of new species, it was their confidence in evolutionary change that resulted in Hugo DeVries's adequate mutation theory. But faith in a guiding principle is no guarantee of its permanent success or of its final truth. Einstein's faith in the aesthetic guiding principle of symmetry in nature has not as yet produced any sign of objects gravitationally repelling each other, even though both repulsion and attraction are found in magnetism and electricity. Nor has a Newtonian or an Einsteinian faith in another, our sixth, guiding principle, a *basic unity of nature,* yet established a field theory unifying gravitational and electromagnetic fields.

Both unity and symmetry are aesthetic guiding principles that date back to Greek natural philosophy. Unity was another of Newton's famous "Rules of Reasoning." In the words of Frank Manuel in *The Religion of Isaac Newton* (1973), "In whatever direction he [Newton] turned, he was searching for a unifying structure. He tried to force everything in the heavens and on earth into a grandiose but tight frame from which the most minute detail could not escape."

John H. Schwartz was expressing a guiding principle of unity—mathematical unity—when he said the following in the *Scientific American* in February 1975: "As a philosophical matter, I am inclined to believe it should be possible to find a complete, consistent and even elegant mathematical description of all the [subatomic] particles and their interaction." There is also an appeal to aesthetics in the word "elegant."

The following extract from *Science Newsletter* (January 4, 1975) further illustrates that symmetry, like unity, still operates as a guiding principle in modern physics, although not dogmatically so:

> Symmetry is an essential concept in physics. Not only because of a philosophic or aesthetic feeling that nature ought to be balanced, but because of theoretical and experimental evidence that in fact nature is symmetrical in many important respects. Three basic symmetrical principles lie at the basis of particle-physics theory: that there are equal amounts of matter and anti-matter in the universe (charge conjugation), that nature makes no distinction between left-handedness and right-handedness (parity), and that a particle going forward in time looks the same as an antiparticle going backward in time (time reversal). Other symmetry principles apply in particle physics and other branches of the science, some of them more easily definable in mathematical than in physical terms.
>
> When the experimental facts are examined, it turns out that symmetries are sometimes broken. A principle that applies in general will be violated in one or two instances. Such symmetry breaking sometimes appears as a kind of nuisance, a blemish on the face of an otherwise beautiful theory. In other

cases a judicious amount of (theoretical) symmetry breaking enables theorists to construct unified field theories that they could not have without it. Symmetry breaking is thus of crucial interest to physicists, and if there is some mechanism that turns it on and off, what it is and how it works would be important to know.

Music has its themes and counterthemes. Guiding principles also can be opposing and recurring as motifs. Heraclitus's principle of a dynamic universe of motion and change is reflected in Galileo's revolutionary concept of inertia (Newton's first law of motion) or Darwin's theory of evolution. Countering this Ionian motif of universal motion and change was the permanency principle of Parmenides holding that "there is nothing new under the sun." Today, one of the fascinations of theoretical physics is its many physical constants such as the gravitational constant G, Planck's constant h, or the absolute value of the speed of light c. Such universal constants are modern sophisticated forms of Parmenides' motif of permanency.

In later chapters, where relevant, we will further discuss such apparently paradoxical guiding principles to nature as continuity versus discontinuity, analysis (into basic components) versus synthesis (into patterns), unity versus duality, determinism versus chance, or symmetry versus asymmetry, which all lie at the heart of modern ideas of matter, energy, and cosmology. Guiding principles, we repeat, have much of intuition, imagination, and aesthetic feeling in their makeup. They are worth examining. They reveal nuch that is subjective, human, and innovative in scientific inquiry. Guiding principles are pretty much guiding faiths that run through history as motifs behind changing conceptual models. There are no basic evidences or proofs for guiding principles, as, for example, simplicity or mathematical harmony. But the principles lead to theories, laws, and hypotheses that can be verified or falsified. The productive polarity of paradoxical guiding principles continues. In our comparative ignorance, nature as yet remains a sphinx.

When Einstein projected non-Euclidean geometry upon a formerly Euclidean universe, the principle of a mathematical universe still remained intact even though the geometric conceptual model changed. Is mathematics a language of *nature* or a language of *science*? The difference constitutes the significance of a guiding principle. Science is a product of man, and man, a product of nature, but what nature is, we are still trying to determine. Meanwhile the universe is treated as mathematical. And, in that connection, is mathematics a metaphor through which we project an equal sign (=) upon the universe? As William James expressed it, "Every philosopher or man of science whose initiative counts for anything in the evolution of thought, has taken his stand on a sort of dumb conviction that the truth must lie in one direction rather than another, and on a sort of preliminary assurance that his notion can be made to work; and has borne his best fruit in trying to make it work."

Interaction obviously is a word that has come into more and more use. It suggests dynamic interrelationships among people and things rather than a static isolation. Nuclear particles interact to form new entities. Atoms interact to form molecules. Molecules form cells; cells form tissues; tissues and organs evolve into ever greater complexity until we reach man and his manifold interactions with nature.

Indeed, we have emphasized in this chapter the human subjective, interactive aspects of the scientific quest for knowledge. Scientists must be seen not as completely detached, independent observers, but as conscious parts of nature, interacting with, even intervening in, other parts of nature whether they are observing, measuring, imagining, explaining, forming and testing hypotheses in self-corrective feedback processes, or technologically changing the face of the earth. We approach our surroundings using everything we have: senses, mind, imagination, and experience, including socially accumulated experience and techniques. The scientist gathers data but he shapes his data through conceptual models and guiding principles much as a sculptor shapes clay. And he projects his tentative models upon nature itself. Much like the artist, the scientist, in a sense, partially creates the order that he seeks. More than that, the scientist forms a system with what he or she observes. This system of interaction necessitates a further look, in chapters to come, at the nature of scientific objectivity; at conceptual system building in science; and at man's place among the hierarchy of innumerable system within systems, systems both inside and surrounding us. Perhaps scientific objectivity and subjectivity themselves are to be understood in such a context of science as human interaction with nature. That is, can a scientist be part of a system and still be really detached from that system?

NOTE

1. Albert Einstein and Leopold Infeld, *The Evolution of Physics* (New York: Simon and Schuster, 1952).

SUGGESTIONS FOR FURTHER READING

Beveridge, W. S. B. *The Art of Scientific Investigation.* New York: Vintage, 1957
Bronowski, Jacob. *The Common Sense of Science.* Cambridge, Mass.: Harvard University Press, 1977.
———. *Origins of Knowledge and Imagination.* New Haven: Yale University Press, 1978.
Hanson, Norwood Russell. *Patterns of Discovery.* New York: Cambridge University Press, 1965.
Holton, Gerald. *Thematic Origins of Scientific Thought.* New York: Cambridge University Press, 1973.

11

Scientific Investigation as Interaction: A New Look at Objectivity

There is no way to disentangle the objective, external regularity from the subjective human form of experiencing, conceptualizing and imaginatively
exploring that regularity at any given historical moment.

M. Markovich, personal interview, 1960

Knowledge is structured in the human mind and imagination. This structuring
is part of a larger process of human interaction with the surroundings. Better
still, we may say that knowledge is gathered by the senses; structured by the
mind; and reinforced by experience, imagination, and intuition. Resulting
conceptual images are often projected back upon nature as if they were nature
itself, although they need to be differentiated from actuality. This image projection also forms part of the larger feedback process of scientific interaction
with the surroundings.

What are the specific aspects of scientific interaction with nature and
how do these aspects form an interactive self-correcting process of inquiry?
Fig. 11.1 suggests, in concise form, answers to these questions for introduc-
tory elaboration in this chapter.

Nature contains both objects and images. There is the massive sun in space
and there are small images of it mirrored on the retinas of observers. The sun
is a fact of nature that in science is assumed to exist independently of observers:
it radiates light, heat, and other energy forms that bathe surrounding space. Such
solar radiations act as signals informing us of the sun's existence; reflected from
various objects, they may strike a sensitive camera, eye, or other receiving
mechanism. This mechanism responds to give size, shape, color, and other
sensory impressions or images of objects. It is clear that observation has three
components: (1) the observer, (2) the observed, and (3) intervening signals.
These components form the basis of a system of interaction, whose elements
are indispensable to each other. See Fig. 11.1 A and B, sections 1, 2, and 3.

169

Fig. 11.1
Scientific Inquiry as Interaction with Nature

A. The Self-Correcting Investigative Process

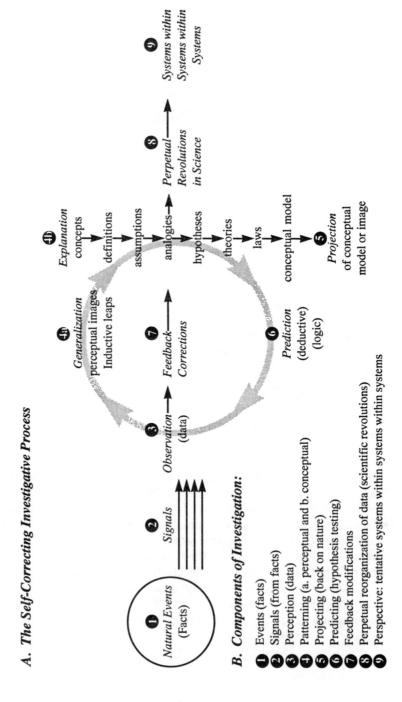

B. Components of Investigation:

1. Events (facts)
2. Signals (from facts)
3. Perception (data)
4. Patterning (a. perceptual and b. conceptual)
5. Projecting (back on nature)
6. Predicting (hypothesis testing)
7. Feedback modifications
8. Perpetual reorganization of data (scientific revolutions)
9. Perspective: tentative systems within systems within systems

The sun's radiation thus does more than provide data from original natural sources or events. The radiation signals help form a system, a continuity of components: object—> signal—> observer when they strike a sensitive eye or camera. And, of course, thanks to the human nervous system and brain, the images or impressions formed were conceptual as well as perceptual (Fig. 11.1, components 4a, 4b, and 5). But objects are not images. Images, perceptual and conceptual, at best can approach but hardly incorporate full reality. And, in breaking through comparative ignorance, science as yet deals primarily with its images rather than directly with objects. Our sensory impression of the sun is that of a simple glowing yellow orb, one angular degree in diameter, which does not give the accurate size, shape, content, elemental colors, internal nuclear processes, and other details of the sun itself. Telescopes, spectroscopes, and other extensions of the senses, of course, do add somewhat to our information. By providing more data, they help achieve the closer correspondence between objects and images that science seeks. But we perceive little of what goes on within the sun. what has kept it glowing for not millions but billions of years? In answer, with the theoretical physicist Hans Bethe, we shape a conceptual model of nuclear fusion in which hydrogen fuses into helium within the sun, emitting tremendous energies in accordance with Einstein's $E = mc^2$ (where E is the energy emitted, m is the mass transformed, and c^2 is the square of the speed of light) . Here, in comparative ignorance, science calls the world to order through conceptual images (e.g., nuclear fusion and Einstein's equation). At best, with our images, which need to be constantly revived, we approach but hardly incorporate full reality. Consequently, we must distinguish natural events, signals, and human images from each other as components of an interactive knowledge process involving an observer and objects that he or she observes. The sun as an independent event or *fact* of nature is to be differentiated from both the *data*-bearing signals it radiates and the sensation and conceptual images of responding human mechanisms as emphasized in Fig. 11.1 A and B. If facts, data, and images (perceptual and conceptual) are not differentiated, confusion results. Facts are the *natural events* emitting signals. *Data* involve sensory apparatus responses to signals whether these be color, sound, smell, taste, pain, a rainbow, or dial readings. Thus, in data we deal more directly with energy signals *from* events rather than directly *with* facts or events themselves. Our information is, at best, partial.

Fig. 11.1 A and B focuses on the data specifics of "observation" at the contact point 3 with incoming signals. Note with the upper curved arrow that the observed specifics of data are inductively generalized and structured through language (concepts, definitions, descriptions) as well as through assumptions and analogies to form hypotheses, theories, laws, and conceptual models 4a and 4b. All these are indispensable to explanation. Movement in such an inductive generalizing process proceeds from the specific to the general, from the particulars of observation to the generalities of language and law. Early observations of the moon first led Newton to the universality of his law of gravitation. Note further in Fig. 11.1.A that "prediction" completes the circle to "observation" (3). Future data are predicted from the generalization (e.g., a hypothesis, theory, or law) to be tested; these predicted data are then checked against later observed data (3). Nineteenth-century astronomers predicted from Newton's law of universal gravitation the presence of the yet unseen planets Uranus, Neptune, and Pluto. Movement here was from the general law to the specific discovery, and therefore represented deductive logic. The whole circle enables a questioning of nature by forming hypotheses about nature and obtaining answers by prediction, feedback, and (often) correction of data (3) or ideas (4) (Fig. 11.1A) via component 7. Certainly we may observe here a follow-through interaction between scientific man and nature from an original contact point of observation (3) in Fig. 11.1 A.

That is, scientists do not just observe; they form ideas and images which they then project back upon nature to explain what they see. In Einstein's words,

> Physical concepts are free creations of the human mind, and are not, however it may seem, uniquely determined by the external world. In our endeavor to understand reality we are somewhat like a man trying to understand the mechanism of a closed watch. He sees the face and the moving hands, even hears its ticking, but he has no way of opening the case. If he is ingenious he may form some picture of a mechanism which could be responsible for all the things he observes, but he may never be quite sure his picture is the only one which could explain his observations. He will never be able to compare his picture with the real mechanism and he cannot even imagine the possibility or the meaning of such a comparison. But he certainly believes that, as his knowledge increases, his picture of reality will become ever simpler and will explain an ever wider range of his sensuous impressions. He may also believe in the existence of the ideal limit of knowledge and that it is approached by the human mind. He may call this ideal limit objective truth.[1]

In sum then, as shown in Fig. 11.1 A and B, all human components of interaction may be telescoped into these five: perceiving (component 3), patterning (component 4), projecting (component 5), predicting (component 6), feedback correction (component 7), and perpetual reorganization (component 8). *Perceiving* involves sensory reception of signals from natural events, at the heart of scientific data. *Patterning* involves formation of data into visual and conceptual images in the human mind and imagination. *Projecting* pertains to casting these creative images or conceptual models out upon nature for unifying explanation of events. *Predicting* enables testing the projected models. *Perpetual reorganization* of data into new conceptual models enables better fit between data and models through repeated prediction feedback and anomalies. Eventually there results a perspective of related systems within systems within systems, or conceptual worlds within worlds within worlds, in nature (component 9). Recognition of such extends insights into the significance of context in understanding nature and in the benefits of a holistic approach that includes recognition of the observer's place in the whole picture. All this will be described in considerable detail in later chapters.

Long before twentieth-century revolutions in physics, there were indications that an observer is not independent of what he observes. Newton discovered through his law of universal gravitation that the weight of an object is not an independent characteristic of that object but the result of the external force of gravity acting upon the object from the earth's center. When you step upon a platform scale, gravity presses you down on the platform to give you a reading that depends not simply on the mass of your own body but on the mass of the earth as well. It also depends inversely upon your distance squared from the earth's center. On the moon's surface, you would weigh less. The moon's attracting mass is less than that of the earth (even though your distance to the moon's center is less than to the earth's center). Thus, your weight fully and definitely involves a system of interaction of two different objects (you and the earth) and is not an independent property of yourself. The earth not only exerts a gravity force upon you, but you exert an equal backpull on the earth. This was easily seen in the case of the moon and the earth (see chapter 9). The earth exerts a gravitational force upon the moon that keeps the moon in orbit; the moon also exerts an equal force back upon the earth that shows itself in ocean tides. These forces of action and reaction do not act upon the same object but upon two separate objects, the moon and the earth. Or upon you and the earth. We start with the matter of weight—supposedly your weight—and end with a system of interaction between you and the earth.

The apparent half dome of the heavens is an image within us of a space framework that does not actually exist. The earth's spherical shape and our own eye characteristics and limitations partly determine what we "see." Besides, we would not "see" that particular half-dome shape of the heavens if light rays did not travel from various points of the horizon and sky to our eyes, and in the particular "straight" or geodesic lines that they do. And if light had a different speed, the stars would not appear in the particular positions and patterns that they do. When we gaze into the heavens, our retinas do not give us a mirror image of where particular objects actually are *now* in space because it has taken light variously up to several billion years to reach us from different stars. Meanwhile the stars themselves have moved on or died. We are looking, therefore, into past time as well as space. Furthermore, we see heavenly objects in a four-dimensional space-time, said Einstein. But our senses alone cannot be trusted in a three-dimensional, let alone a four-dimensional, world. This will be more fully discussed in a later chapter on Einstein's theory of relativity.

Ancient man consciously patterned groups of stars into constellations and a zodiacal belt. The observed relative positions of stars suggested certain patterns to these observers. But why were all the constellations named after living things (except one, Libra)? These patterns were certainly creations and projections of man's mind, experience, and imagination. Involved was a system of interaction between the stars and men, with light signals traveling between the two in a pattern already observed in Fig. 11.1: natural events (facts) —➤ signals (from facts) —➤ perception (sensory data) —➤ patterning (e.g., constellations) —➤ projecting (back upon nature).

Early earth-centered astronomers said "the sun rises" because that is what they observed. We still use those words today even though, as Copernicans, we believe that from a rotating earth, the sun merely *appears* to rise in relative motion. A *supposedly* detached impersonal observation had it that "the sun rises." And yet if the observer forms a system with the sun that he observes, the observer could actually be moving instead of the sun. Motion is relative.

Later, when Ptolemaic astronomers projected celestial spheres upon the heavens, they used geometry to shape astronomy. By projecting spheres first as conceptual frames of reference of moving points of light and then as majestic rotating star-bearing "clockworks" (Kepler's word), astronomers were working not as detached observers, but in interaction with nature. In an interpretative sense, they were intervening in nature, "calling the world to order." Science today, through its equations, projects equal signs or else expanding-universe models onto the heavens. Einstein, in assuming order in the universe, created order, a four-dimensional space-time universe.

OBJECTIVITY AND INTERACTION

Subjectivity in science was discussed in previous chapters in terms of imagination; intuition; guiding principles; motifs; models; metaphors; hypothesis formation; and other human, social, and professional encapsulations which bind us.

Let us now consider the other side of the same coin: objectivity. Are scientists merely spectators or interactors in nature? Upon the answer to this query of Werner Karl Heisenberg (1901–1976), hangs a problem of scientific objectivity. Objectivity, which has long been assumed in science, implies detachment of the observer from what he or she observes. The nature of this detachment can be inferred from the following definition in *Webster's New Collegiate Dictionary* (1984): "having the status of or constituting an object existing independently of mind, belonging to the sensible world and being observable or verifiable especially by scientific methods; emphasizing or expressing the nature of reality as it is apart from personal reflections or feelings; expressing or involving the use of facts without distortion by personal feelings or prejudices; relating to or being methods that eliminate the subjective by limiting choices to fixed alternatives requiring a minimum of creative interpretation." Emotional bias is minimized in science through mathematical language and logic as well as through operational definitions and hypothesis testing as emphasized in former chapters. But is detachment really possible in observing and structuring data? Human physiology, optical theory, relativity, quantum theory, and cybernetics resoundingly answer no. The observer, we repeat, forms a system with what he observes: with his senses, mind, imagination, and measurements, he interacts and even intervenes in his investigation. How can he be part of a system and still be detached from it? The value an observer obtains, for example, in measuring the length of an object depends on the observer's speed relative to the object (Einstein). Color and other sensations are responses to light and other signals of events. A rainbow is not an independent entity but a phenomenon of interaction involving the sun's rays, moisture, and an eye or other receiving mechanism properly positioned in respect to each other. With his apparatus the scientist intervenes in what he observes and often seriously influences the character of his results: When diffraction gratings are used in light experiments, light appears wave-like; when electric "eyes" (photoelectric cells) are used, light appears particle-like. And the very light or other radiation we use to observe objects (large or small), by impact disturbs the motion and position of these objects (Heisenberg).

Heisenberg seemingly answers his own question as to whether "scientists are detached spectators or interactors in scientific inquiry." He says, "The object of research is no longer nature in itself, but rather nature exposed to man's questioning and to this extent man here also meets himself." "Man meets himself," of course, with the conceptual models and images that he projects upon nature and through which he structures his data: What does this do to detachment? To scientific objectivity? And what are its implications for the very nature of scientific knowledge itself? For example, what do we have when we have a law in science? Is it man-made, at least in part, or an absolute principle of nature? Is a scientific law discovered, invented, or both?

How could Newton's law of universal gravitation reign supreme for over two centuries and then be supplanted as a universal law of nature by Einstein's general theory of relativity? Just what does that do to the laws of nature and to the nature of scientific knowledge? Isn't "natural law" to be understood in terms of human *interaction* with nature rather than simply in terms of independent properties of nature itself—especially under the encapsulating conditions of very limited knowledge, experience, and human sensibilities?

And isn't objectivity to be somehow redefined in terms of the observer's interaction with nature rather than of observer detachment and the independent properties of things? We have been suggesting here that if knowledge, with its data, principles, models, practices, and techniques, is structured in the human mind and imagination, the following considerations are feasible:

1. The structuring is part of a larger process that includes an open-ended system of events, signals, perception, conceptual patterning, projection, prediction, perpetual reorganization of ideas, and broadening of perspectives. We form images of things that we project back upon the world. It is these projective images that we ordinarily deal with rather than things or people themselves. At least in science we test and modify these images. It is in these projected images that, in part, "we meet ourselves."

2. The above process is open-ended and involves conceptual, imaginative system building and system projection upon nature.

3. Operating upon incoming data or signals from natural events are interacting senses, mind, imagination, intuition, muscles, and apparatus.

4. Sensations, ideas, assumptions, images, models, and beliefs lie at the heart of knowledge as product ingredients of scientific interaction.

5. Natural laws, theories, and other scientific concepts and models (all human creations) are to be reexamined and redefined in terms of human interaction with nature rather than in ordinarily simplistic terms of the properties of nature itself. This is well illustrated by Newton's interactive concept of weight or Einstein's interactive, relativistic concepts of mass, energy, space, time, and force.

6. All this is in line with basic assumptions of nature existing before human investigators, with human beings evolving from nature, and with science evolving in human interaction with surrounding nature.

7. The world is so constructed that when creatures of our particular senses, sensibilities, and makeup emerge, the world appears as our senses and minds pattern it. That mankind has survived means that man, in his interactions with nature has in the main been in tune with nature thus far. This would be in line with Darwin's principle of natural selection. (Perhaps at this point lagging social intelligence had better catch up with scientific and technological advances!)

To further emphasize that scientists are "interactors rather than spectators of nature," let us recapitulate even further.

SOURCES, SIGNALS, AND SIGHTS

The sun shines and gives forth light, heat, and other radiation forms. The "shining sun" is a natural event with its own interior processes. Radiation emitted by the sun may be considered signals of the event and its processes. Such natural events and their signals are independent of observers and may be considered to be facts. They exist with or without observers: that is a basic assumption of science. Such events are *objects* of observation about which we observers form *images*. We depend upon radiation or other signals between us and objects for our sensations, impressions, and ideas of the objects. We deal directly with these as signals rather than the objects themselves through our very limited senses and data. More than that, what is directly involved in perception is *interaction*, say, between light signals and our limited eye mechanisms. Remember that our eyes respond to only one octave (called light radiation) in a very broad spectrum of over sixty octaves of electromagnetic signals that leave the sun. Color is a sensory response to light waves. Rainbow color sensations and our perceptual images of objects are limited to this narrow one-octave range of interaction. The danger, as in all encapsulation, is in treating partial pictures as the entire picture, object, or event. Let us, therefore, distinguish between *facts* and *data*, between original actual *events* (e.g., the sun and its processes) and *signals* from such events (e.g., light radiation). Data form at a sensory level of interaction. And telescopes, microscopes, and man-made instruments with dials are basically extensions of the human receiving mechanism. Data can be factual in part, but not necessarily so. The grass is everything but the green in which we say we "see" it. Color is a sensation of *interaction* that does not really depict the original object as is. Color awareness and sight are limited physiological responses in reaction to limiting light stimuli. And strangely enough, the grass itself contains every color but the greenness that we experience. (This will be explained in a later chapter.) Again, in nature there are objects and there are images. Whether on camera film or the eye's retina, a yellowish orb of the sun appears as an inverted color image. The sun itself, as a huge, dynamic, natural object independent of man, is to be differentiated from man's sensory image of it. Hans Bethe conceptually visualized the sun as a gigantic natural thermonuclear furnace that fused hydrogen atoms into helium atoms and emitted a seemingly inexhaustible array of radiant energy signals. But Bethe's conceptual image is a long way from our sensory image of a yellow orb one degree in diameter. His conceptual image (excellent theory) is to be differentiated from the sensory image (radiation data), and both are to be differentiated from actual events (facts) within the sun to be further investigated. Light, sight, and thus color are limited joint products of interaction between the sun's rays and eye mechanisms.

Whatever the *events* occurring within the sun itself, the sun's rays are *signals* of these events perceived by the eye. The character of color sensations that we experience depends, among all else, upon actual events in the sun, the nature of the signals emitted, and the natural selectivity of the eye.

Thus, in general, we deal at best with our images or partial pictures of people, things and events: we are encapsulated. These partial pictures, images, conceptual systems, or whatever we choose to call them are selected and shaped by our sensory, conceptual, and imaginative apparatus.[2] We shall see later that the nervous system, impressions, language development, past experience, and sociocultural conditioning are also all involved in these images and conceptual models with which we approach nature. At this point, we merely allude once more to the illustrations of the blind men and the elephant, the seeming half dome of the heavens, the Babylonian zodiacal belt, the Eudoxean spheres, Hipparchean epicycles, and the Keplerian elliptical orbits or our present concepts of an expanding universe that supposedly originated from a Big Bang. Incoming sensory data, conceptual model building, and model projection back upon nature are all involved. Conceptual models relate data, definitions, classifications, assumptions, ad hoc hypotheses, intuition, imagination, and eventually theories and laws. The sequence, to repeat, generally is: an *event* (*fact* of nature), *signals* from the event, *perception* by a receiving mechanism (data), *patterning* (perceptual and conceptual), *projection* of the pattern back upon nature, *prediction* of future observation, feedback and *perpetual reorganization,* and testing of ideas with the eventual development of *interrelated system within systems within systems.* This entire process is primarily one of interaction between scientific man and nature in an open-ended rise, development, decline, and eclipse of conceptual models and associated practices (Kuhn's paradigms). Operating in the reaction are human senses, mind, imagination, feelings, intuition, muscles, and apparatus.

Again, we have seen that the knowledge process involves: (1) actual independent events; (2) their signals to us; (3) our observations; (4) joint formations of outside nature and man, such as a rainbow or other sensations; and (5) our interpretations, projections, predictions, and applications. There are many difficulties in attempting to distinguish among these components of a knowledge system. First, we are only conscious components of the systems, attempting to fathom the whole with very limited senses, sensibilities, and experience, and with a host of encapsulations, individual and social. We shall also see that there is no observation without a frame of reference and no frame of reference without interpretation. Since frames of reference bridge observation and interpretation, there is no hard and fast rule as to where observation ends and interpretation begins. As in a color spectrum, where to draw dividing lines is a difficult proposition.

Particularly challenging are events, such as a rainbow, that are joint products of the observer and nature. The sun, moisture-laden clouds, and light rays exist independently of an observing eye, but a rainbow as such does not exist until rays from the sun are reflected from moisture to the eye. The eye and brain, as conscious parts of nature, are not passive mirrors of outside events but actually create, with external nature, joint components of knowledge itself. As elsewhere in nature, interacting parts create qualitatively new phenomena or patterns, in this case, a rainbow. Disturbances in the air known as sound waves do not sound until they impinge on an ear. Solar radiation becomes body heat only when absorbed by tissue molecules.

We may say that the subjective enters into the objective, and the objective includes the subjective. We shape what we see in human interaction with surroundings in survival and in search for knowledge. If the observer forms a system with what he observes, he brings his encapsulating selectivity and limitations to what he observes. That is, he creates and intervenes as he observes. Observation is dynamic and interactive, not passive.

Thus, beneath the surface of scientific objectivity lie many complexities in sensory and mental selectivity, in roles of separate brain hemispheres, in other limitations of observer interaction with nature, in comparative human ignorance, in the role of language, social conditioning, and many other encapsulations. Objectivity, therefore, is more fully understood at different levels of complexity in our interactions with nature. All this will be pursued in more detail in succeeding chapters.

NOTES

1. Albert Einstein and Leopold Infeld, *The Evolution of Physics* (New York: Simon and Schuster, 1952).
2. See again Fig. 11.1 and the poem that appears on p. 5 of this volume.

SUGGESTIONS FOR FURTHER READING

Kuhn, Thomas S. *The Structure of Scientific Revolutions.* Chicago: University of Chicago Press, 1970.

Perlman, James S. *The Atom and the Universe.* Belmont, Calif.: Wadsworth Publishing Co., 1970.

12

Perceiving as Interaction: Sources, Signals, and Senses

There is more to seeing than what meets the eyeball.
N. R. Hanson, *Patterns of Discovery,* 1958

MYSTERY OF THE RAINBOW: LIGHT, SIGHT, AND COLOR

The rainbow in all its poetic splendor has intrigued man ever since he could first observe and speculate. It has explanations in mythology, it enters into the Bible, it fascinated Greek natural philosophers, and it allured the mathematically minded Newton. Let the rainbow now illustrate for us the interactive character of perceiving, a process that involves vibratory sources, signals, and senses.

There are three prerequisites for seeing or hearing: (1) a vibrating source, (2) an emitted signal, and (3) a receiving mechanism. There is no vision without light from a source striking an eye. No sound can occur without sound waves striking an ear.

What is a rainbow? Where do its colors come from? Its size and shape? Adequate answers depend upon satisfactory solutions to other questions such as the following:[1]

1. Why does sunlight leave shadows behind trees (unlike floodwaters, which completely surround obstacles)?

2. Why does an oar partially submerged in water appear bent?

3. Why does the rising or setting sun appear more red than the midday sun?

4. If you extend a right hand to your mirror image, why does the image extend a left hand back?

5. Does a color exist if no eye is there to see it?

181

These five questions involve relationships among light, sight, and color that took man almost his whole history to understand. A synthesis was eventually made by Newton (see chapter 15) in a superb model of experimentation that answered the question of rainbow colors.

Questions such as the above also lead to the awareness that knowledge is a product of interaction between man and his surroundings rather than the acquisition of an independent, completely detached observer. That is, the observer forms a natural system with what he observes. Again, in N. R. Hanson's words, "There is more to seeing than what meets the eyeball."[2]

However, it was Greek and Roman natural philosophers who, over two thousand years ago, laid the foundations for Newton and for us today. In 300 B.C.E., Euclid observed that "light travels in straight lines called rays." In so doing, this great thinker was applying mathematical techniques to nature by arbitrarily reducing light to rays. In so doing, he was calling the world of light to order and, incidentally, explaining shadows. Water waves bend around obstacles; light rays traveling in straight lines do not, and thereby leave shadows.

Hero of Alexandria (first century C.E.) experimentally established Euclid's hypothesis by "bouncing" narrow beams of light off plane mirrors. The path of the beams, although broken by the mirror, was straight in approaching and leaving the mirror. But even more significantly, Hero found that the angle at which a beam approaches the mirror always equals the angle at which it leaves the mirror (Fig. 12.1). The perpendicular dashed line *PO* drawn to the mirror surface in the illustration helps form the two equal angles. (The incident ray, the reflected ray, and the perpendicular all lie on the same plane.)

Light reflected from mirrors or smooth polished surfaces produces images. Your image in a plane mirror (Fig. 12.2) results from the reflection of light rays off the mirror's surface. Since your eye does not catch the ray's change in direction at the mirror's surface, you see yourself behind the mirror. And because of the rays' change of direction, the left hand of your image appears to meet your outstretched right hand. We cannot always trust our senses alone; we intuitively react to light rays as if they were traveling in unbroken straight lines. The direction that "counts" is that from which the ray enters the eye from the mirror. Again, *in "seeing," the observer forms a system with what he observes in which illusions may result.*

But light rays are not always merely reflected when they strike the surface of objects; they may also be absorbed or transmitted. Some rays are reflected from water surfaces to give a mirror-like effect; others may be absorbed by muddy water. Oars are seen in water when light rays penetrating the water are reflected back by the oars. Aristotle was among the first to describe correctly the broken appearance of oars in water. But it was the astronomer Ptolemy who first explained the broken appearance through the concept of *refraction,* or the bending of light rays (see Fig. 12.3). In the fifth volume of his *Optica,* or treatise on optics, Ptolemy writes:

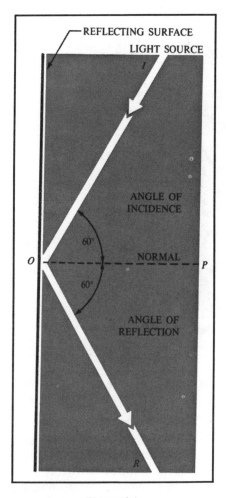

Figure 12.1

Reflection of light. The angle of incidence equals the angle of reflection ($\angle i = \angle r$).

Visual rays may be altered in two ways: (1) by *reflection,* i.e., rebounding from objects called mirrors, which do not permit penetration by the visual ray, and (2) by *refraction,* in the case of media which permit of penetration and [are transparent]. It has been shown . . . (1) that a visual ray proceeds along a straight line and may be naturally bent only at a surface between two media of different densities; (2) that the bending takes place not only in the passage from rarer and finer media to denser . . . but also in the passage from a denser medium to a rarer; and (3) that this type of bending does not

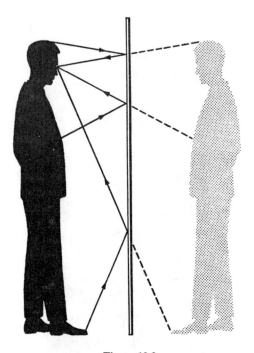

Figure 12.2

Your image results from the reflection of light rays at a mirror surface. Since your eye does not catch the rays' change in direction, you see yourself behind the mirror.

take place at equal angles but that the angles, as measured from the perpendicular, have a definite quantitative relationship.[3] (Italics added)

A century before Ptolemy, Seneca (3 B.C.E.–65 C.E.), the Roman natural philosopher (and adviser to Nero) seriously reflected upon the rainbow. He posed questions that are still cogent today: What causes the rainbow? Why does it appear after rain? From where arises the variety of colors? Why the semicircular shape and the large size? Why is the bow seldom found during midday in the summer?

In his *Naturales Quaestiones* (*Natural Questions*), Seneca uses a mirror analogy in establishing the origin of the rainbow in a *positional relationship between the sun, clouds, and an observer.* The clouds become an imperfect mirror giving a distorted image of the sun seen by a properly placed observer. The rainbow exists as an image and not as an independent entity in the heavens. Seneca's mirror analogy is incomplete and even erroneous in explaining origins, color effects, shape, size, and seasonal variations of rainbows; but he

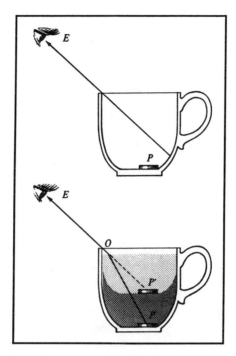

Figure 12.3

Because of the bending of light rays, the unseen coin appears when water is poured into the cup.

made a definite contribution by establishing that for a rainbow, three interrelated conditions are necessary: (1) the presence of water drops in a cloud or mist, (2) the sun shining on the drop, and (3) an observer with his back to the sun and his eyes toward the mist or clouds. For our purposes here, Seneca's observation involves a system and its interrelationships of necessary components, including an observer.

Newton solved the mystery of the rainbow's colors. Granted, it took sixteen centuries and someone with the stature of a Newton to come up with today's basic answers, but how far would Newton have gone with his analysis of sunlight without the previous refraction ideas of Ptolemy and Snell's mathematical law of refraction?[4] Also, Archbishop Marco Antonio de Dominis (1566–1624) and René Descartes had correctly applied principles of refraction and reflection in explaining the rainbow before Newton. But neither Dominis nor Descartes could adequately account for how refraction results in color. Newton became interested in the fringes of color that blur telescope images. In systematically investigating such color phenomena with prisms, he determined that white light is a blend of colors, and that prisms, lenses, and raindrops act as refraction agents for separating the component colors of white light. As Newton reported in a letter to the Royal Society:

> Having darkened my chamber and made a small hole in my window-shuts to let a convenient quantity of the Sun's light, I placed my Prisme at his entrance, that it might be thereby refracted to the opposite wall. It was at first a very pleasing divertisement to view the vivid and intense colours produced thereby, but after a while applying myself to consider them more circumspectly, I became surprised to see them in an oblong form, which, according to the received laws of Refraction, I expected should have been *circular*.
>
> They were terminated at the sides with straight lines, but at the ends, the decay of light was so gradual, that it was difficult to determine justly what was their figure; yet they seemed *semicircular*.[5] (Italics added)

Newton named the band of colors a *spectrum* and established seven colors by analogy with seven notes of a musical scale: red, orange, yellow, green, blue, indigo, and violet (Fig. 12.4). It was not the colors appearing in a prism in gradual transition from one to another that surprised Newton; men had noticed rainbow effects in glass even long before Seneca. What intrigued Newton was that these colors formed an oval with semicircular ends. Why did this shape result when the light streamed upon the prism through a round hole?

Through a series of ingenious experiments with prisms, Newton established that white light is a blend of colors. Figures 12.4 and 5 are a simplified representation of one of the experiments. The first prism merely disperses, that is, separates the colors composing the white light by bending each at a different angle. The second prism reunites the colors into white light by bending them variously toward the thicker part of the glass.

Newton's prism experiments show that light bends or refracts increasingly as we move from the red end of the spectrum to the violet. The blue component of sunlight refracts in the prism far more than the red. If by the same token we assume that blue light is scattered far more than red by dust or other fine particles in the atmosphere, we can explain why the sun appears redder when it is setting than it does at noon. At sunrise or sunset, with the sun at the horizon, sunlight passes through a larger mass of atmosphere to reach the eye of the observer than it does at other times of the day. The red component scatters less than the blue, and more red comes through directly to the eye. Objects at the horizon opposite to the sun take on a bluish cast because of the much greater scattering of the light. Also, it is the scattering by the atmosphere of the blue in sunlight at any time of the day that gives the heavens their general blue color. Balloon and rocket observations above the earth's atmosphere and scattering show the sun as a glaring white disc in a black sky.

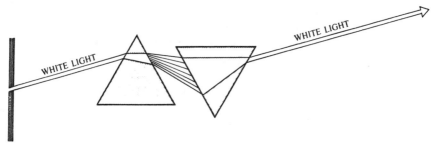

Figure 12.4

Dispersion of light. White light is a blend of colors. The first prism merely separates the colors already there by bending each at a different angle. The second prism reunites the colors into white by bending them variously toward the thicker part of the glass.

COLOR BY SELECTIVE REFLECTION

Greenness does not reside in grass itself but is a sensation in your eye. By the principle of selective reflection first recognized by Newton, the color of an *opaque* object depends upon the color of the light that it *reflects* to the eye. White light is a blend of all the colors of the rainbow. Grass absorbs most light radiations of the sun and reflects primarily the green radiation. What determines the green sensation is not the radiation that the grass absorbs, but the remaining radiation that the grass reflects to your eyes. The color in grass is everything but the green you see. Similarly, the red is not in a rose that reflects primarily red radiation. Minor amounts of a few other color radiations along with red give a sensation of a shade of red rather than a pure red.

In the study of heat, it is necessary to differentiate between heat radiation and heat itself. Heat radiation from the sun becomes heat when absorbed by molecules. It is the same with light; color radiations from an object become color sensations after entering an eye mechanism. Color is an interaction, not a property. Just as, thanks to Newton, your weight involves an interaction between you and the earth, the red that you "see" is an interaction among light radiations, the rose, and you. A pure red object appears black under a blue, green, or pure yellow light that furnishes no red radiation for reflection. And pure green grass appears black when under a blue, red, or yellow light not containing green radiation to reflect to the eye. An object reflecting all radiations of white light quite equally appears white—for example, a page of this book. An object that absorbs all color radiations, and therefore reflects none, appears black—the print on this page, for example.

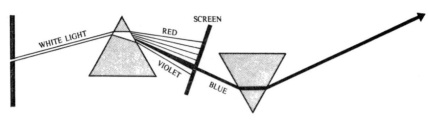

Figure 12.5

White light as a blend. A prism does not change the basic nature of light; rays of a given color (blue) passing through the screen and second prism remain the same (blue).

Newton also recognized a principle of selective color by transmission. In this case, the color of an object depends upon the color of light it *transmits* rather than reflects. Red glass transmits primarily red radiation and absorbs other color radiations. Ordinary window glass is colorless because it transmits all components of white light.

COLOR BY ADDITION

Color is a sensation in response to light radiation. Without an eye there is no color, only light radiation, just as without molecules there is no heat, just heat radiation. Without a world of objects, there is nothing to view, nothing to reflect or absorb radiation. The sun, the eye, and a world of form and color go together. In his *Opticks* (1704), Newton included a "Query" about color vision: "May not the harmony and discord of Colours arise from the proportions of the Vibrations propagated through the fibres of the optick Nerve into the Brain as the harmony and discord of Sounds arise from the proportions of the Vibrations of the Air?" Newton was concerned with the specific roles of the eyes, nerves, and brain in color vision. The color of an object, like its weight, was not to be understood as an absolute property of the object, but as an interaction involving an organism, an outside object, and light.

By 1801 Thomas Young (1773–1829), an English physician-turned physicist, had applied himself to the subjective, or human, aspect of color. An ordinary person can distinguish about one hundred thirty separate hues of the color spectrum. Young did not believe the retina has a separate photoreceptor for each hue that could be recognized and that there would be enough duplications of one hundred thirty receptors across the retina to respond to rays wherever they fell. Young believed that if colored lights could be blended on a screen, they could be blended on the retina. He settled upon three basic color receptors, each sensitive to a specific color: red, green, and blue. These three colors were *primary* colors. All other colors and shades were blends of these three in various proportions and combinations. The idea can be experimentally verified by rapidly rotating a disc with separate colors or by focusing beams of different colors on the same spot. Therefore, only three different color receptors, one for each of the three primary colors, should handle light of any hue entering the eye.

Young's theory was not taken seriously at the time. About fifty years later, the German physicist-physiologist Hermann von Helmholtz (1821–1894) and James Maxwell revived the three-color theory. Helmholtz made an important modification that added his name to the Young theory. He suggested that any of the receptors are activated mostly by one characteristic color radiation and to a much lesser extent by the other two. Pure red light activates the red-sensitive receptors strongly and the other two types of receptors weakly to give a sensation of red. Pure yellow light stimulates the red and green receptors considerably and the blue slightly to give a sensation of yellow.

In summary, the Young-Helmholtz theory offers a process of color by addition. Light of the three primary colors can be added to give any other colored light. Three types of nerve receptors in the retina respond to the primary components in the entering radiation. The brain provides us a sensation of the blend.

Significant here is that the three *primary* colors are *not* an absolute characteristic of nature but are partly determined by the existence of three types of nerve receptors in the retina of the observer. Again, an interaction product!

Because color is a sensation, considerable confusion exists. But this is overcome by differentiating between light radiation that enters the eye and the color sensations that result when color receptors are activated in the retina. A great deal of confusion about color also results if the color of light is not distinguished from that of paints or pigments. The mixing of different colors of light is an additive process; the mixing of paints is a subtractive process. Focusing blue and yellow lights together gives *white* light; mixing blue and yellow paints results in *green* paint. Blue light stimulates primarily the blue receptors, yellow light activates the red and green receptors, since each color receptor is somewhat sensitive to spectral colors on each side of its main color. (Yellow is close between red and green in the color spectrum.) Consequently, blue and yellow light together stimulate all three types of receptors to give a white light sensation. Two colors of light that combine to give white are called *complementary colors*. Other pairs of complementary colors are red and blue-green, orange and green-blue, and green and purple. Complementary colors can be demonstrated by rotating color discs or by focusing spotlights.

Paints or pigments, when mixed, are subtractive because they are substances rather than light. A blue paint absorbs certain components of white light, therefore subtracting from it; a yellow paint absorbs other components from white light. When blue and yellow paint are mixed, together they subtract more of the white light than either singly. In fact, together they absorb all components from white light except that component reflected and seen as green.

These details on light, sight, and color should be sufficient as a first basic consideration that knowledge is a product of interaction between man and his surroundings rather than the acquisition of an independent, completely detached observer. Color as data-forming sensation is a case in point.

SOUND AS INTERACTION

Another quick example of the interaction in observation may be in order. For "observing," i.e., for acquiring sensory information, we can hear, taste, smell, and feel as well as see. The eyes are not our only sensory points of contact with surroundings through which we interact and form systems for information. For example, consider sound. If the wind violently slams a door, there is no sound without an ear to hear it. There are only a vibrating door (and walls) and sound waves—air disturbances set up by the vibrations—not sound as such. For actual sound—a human or animal sensation—there must be a vibrating object (the door), a transmitting medium (e.g., air), and a receiving mechanism (an ear)—all three of which form a system of related parts. As part of a larger system, the ear (and brain) cannot be totally detached in acquiring information. It is a contributor which helps to form the incoming sound wave signals into the sensation or experience of sound. Without an ear to hear there are only sound waves silently passing by, as deaf people know.

The famous Doppler effect is an excellent example of interaction in the sensation of sound. Most of us have experienced the *increase* in pitch we hear as a locomotive whistle approaches. How is that possible if on the locomotive the pitch remains the same? The pitch of any sound *as heard* depends on the number of sound waves reaching the ear per second. As the locomotive approaches and the distance to us decreases, the ear receives from the locomotive source an ever greater number of sound waves per second. This, therefore, raises the pitch. When the train passes and the distance to it increases, fewer sound waves reach the ear per second, and the pitch is lowered accordingly. Once again, the source (locomotive whistle), the signal (sound waves), and the receiving mechanism (ear) form a system that determines the result (changing pitch, in this case). All parts of the system—moving, vibrating whistle, air waves (disturbed air molecules), and affected ear drums—all count, not just the whistle. The process, again, is one of interaction.

THE EAR AS A RESPONDING MECHANISM IN SENSORY PERCEPTION

The ear is a mechanism for transforming incoming air vibrations into sound. This mechanism has three sections called the outer, the middle, and the inner ear. The outer ear starts with the exterior trumpet-like appendage that converges air vibrations into a one-inch canal ending in an eardrum less than a half inch across. Air vibrations, even from a whisper, press against the eardrum to displace it much like a drumstick against a drum head. The eardrum is connected to three tiny bones in the middle ear called the anvil, hammer, and stirrup. Very slight displacements of the eardrum are amplified by these three tiny bones. These amplified movements of the eardrum are transmitted to the inner ear through an oval window connected to the stirrup. The inner ear resembles a cavern filled with a watery fluid containing a snail-shaped cochlea. The twisting interior of the cochlea is lined with thousands of microscopic hair-like nerve cells. Like tuning forks of different sizes, each nerve cell responds to a different vibration transmitted to the oval window and fluid by the stirrup. The waving response of the hair cell in the fluid starts a minute electrical impulse into the auditory nerve. Thousands of such electrical signals are fed into the brain, where they become the sound sensations we experience.

In accordance with sizes of our responding hair cells, we have a hearing range of from about 20 to 20,000 cycles. As in the case of light, sight, and color, our sensory awareness and our ideas are of a limited, selective, relative character. That is, we are limited to the particular receiving mechanisms of the sensory and mental apparatus that we possess. Reality and the universe are sending out their signals in many different ways, some known and some unknown to us. Waves would be one example. We catch only those messages or waves . for which we have receiving mechanisms. For, example, by virtue of our eyes and an accompanying nervous system, we receive waves in a range of from about 3,500 to 7,000 angstroms to give us light, sight, and color. The range of frequencies involved here is a very narrow one in a very broad band of electromagnetic waves including, for example, ultraviolet, X-rays, gamma rays, infrared, and even radio waves. Yet our eyes are not delicate enough instruments to catch any wavelength longer than red light or shorter than violet, just as we are not aware of sound waves below or above a certain pitch or frequency. Yet these waves exist, as we know from our technology. Japanese experts in earthquake phenomena have reported a sensitivity of animals to earthquake waves that human beings would ordinarily be unaware of except through instruments. Radio waves are constantly passing all around us; we are aware of them only when we set up the proper receiving mechanisms. Man has extended his senses through such tools as the radio and seismograph, but the number and diversity of these receiving mechanisms are as yet small. They involve and select only the particular types of messages and knowledge for which they are designed. There should be many more to come which will give us much more precision and information. Meanwhile, we form concepts based upon limited and partial knowledge, concepts that only begin to approach reality.

CONCLUSION

We have seen that *observation* has at least three main components that form a system of interaction: an observer, objects or events observed, and emitted signals in between. The rainbow illustrates observation as interaction. Light rays (signals) from the sun (source) may strike a cloud or prism and be refracted, reflected, and dispersed back to an eye (receiving mechanism) to form a rainbow. The rainbow becomes a synthesized sensation, a joint product of an interaction that involves the sun (event source), light rays (signals), moisture, and an eye mechanism. The sun emits a yellow blend of various light ray frequencies or signals. Clouds separate and disperse the various frequencies. The eye responds to the separate frequencies in projected sensations of corresponding colors. Without an eye to see, there are no colors, no rainbow, just unseen energy waves dispersed in space.

Again, without an ear to hear as a receiving mechanism, there is no sound of a falling tree, only sound waves from the falling tree's vibrations disturbing the surrounding air. The sensation of sound, like color, is a product of an observer-observed system of interaction, requiring the main components of vibrating source, sound waves (signals) in air or another medium, and an ear (receiving mechanism).

In chapter 15 we will trace through the ages what was involved in changing rainbow conceptual models based upon changing observer-observed interactions.

NOTES

1. James S. Perlman, *The Atom and the Universe* (Belmont, Calif.: Wadsworth Publishing Co., 1970), ch. 1 7.
2. Norwood Russel Hanson, *Patterns of Discovery* (Cambridge: Cambridge University Press, 1958), p. 7.
3. See the Ptolemaic Table of Refractions in Perlman, *The Atom and the Universe*, p. 354.
4. See Snell's mathematical law of refraction in Perlman, *The Atom and the Universe*, p. 5.
5. Isaac Newton, *Opticks* (New York: Dover, 1952). This Dover edition is based on the fourth edition of the *Opticks* (London: G. Bell, 1730), with an introduction by E. J. Whitaker.

SUGGESTIONS FOR FURTHER READING

Hirst, R. J. ed. *Perception and the External World.* New York: Macmillan, 1965.
Life Science Library. *Light and Vision.* New York: Time, Inc., 1966.
Ludel, Jacqueline. *Introduction to Sensory Processes.* San Francisco: W. H. Freeman, 1976.
Scientific American's Readings. *Recent Progress in Perception.* San Francisco: W. H. Freeman, 1976.

13

Problems of Objectivity:
The Subjective, and Selective, and the Encapsulating

A father said to his double-seeing son, "Son, you see two instead of one."
"How can that be?" the boy replied. "If I were, there would seem to be four
moons up there in place of two."

Indries Shah, *Caravan of Dreams*, 1968

Man has closed himself up till he sees all things through the narrow chinks
of his cavern.

William Blake, *Songs of Experience*, 1794

Previous chapters suggested serious complexities in scientific objectivity. In
chapter 10 we saw science's subjective, human dimension. *Testing* hypotheses is an objective process: a successful hypothesis or theory must have its predicted results matched by observed results. *Forming* hypotheses, on the other
hand, is generally highly subjective: Motifs, models, and metaphors behind
hypothesis formation are products of human minds and imagination.

Chapter 11 proposed a new look at scientific objectivity by emphasizing
scientific investigation as human interaction with nature. This chapter further
distinguished between subjective and externally objective aspects of the interactive investigative process, and suggested that objectivity in the fullest sense
of the word may be better understood in terms of the total interactive process
and its components (Fig. 11.1). Subjective, human elements exist in the
broader reality of the interactive process and do shape what is seen and
thought. A more complete objectivity would recognize the subjective as part
of itself in this interaction and would work from there to modify definitions,
concepts, theories, laws, and models in terms of interaction. Objectivity itself
as a concept would also then be more fully operational; it would thereby
afford additional insight into revolutionary changes in scientific conceptual
models, ideas, and practices through the ages as, for example, revolutionary

changes in models of the cosmos from Aristotle to Einstein. The way may then be open for anticipating future changes of ideas.

Once again, a rainbow does not exist without the eye of a beholder properly positioned to receive the sun's light rays reflected and refracted from clouds or other moisture. Rainbows are joint products generally of sunlight, clouds and eyes (or other receiving mechanisms). As Alfred North White partly expressed it,

> Nature gets [full] credit which in truth should be reserved [at least in part] for ourselves: the rose for its scent, the nightingale for his song, and the sun for its radiance. The poets are . . . mistaken. They should address their lyrics to themselves and should turn them into odes of self-congratulation on the excellence of the human mind [and feeling]. Nature, by itself, is a dull affair, soundless, scentless, colorless, merely the marrying of materials, endlessly, meaninglessly.[1]

Mankind, however, as a conscious product of nature, can be creative in interaction with the rest of nature (unless or until, of course, man destroys his surroundings and himself). However, the world is so constructed that when creatures of our particular senses and sensibilities arise, things appear to us in the particular way they do.

Chapter 12 provided more details on rainbow phenomena and the expansion of ideas about light, sight, and color. These details hopefully afforded increased insights into interrelationships between the physiological and mental makeup of observers and what they observed as knowledge increased. In this chapter we will emphasize the selective and the encapsulating in our own sensory and mental makeup; our experiences; our language abilities; and our professional, sociocultural conditioning. Our aim is to denote the limiting and subjective as well as creative effects of all these considerations.

SELECTIVITY

> I soon learned to scent out that which was able to lead to fundamentals and to turn aside from everything else, from the multitude of things which clutter up the mind and direct it from the essential.
>
> Albert Einstein, 1952[2]

Selectivity in the knowledge process exists at various levels: sensory, neurological, psychological, intellectual, and sociocultural. There is, for example, an intellectual level of basic assumptions, forms of logic, conceptual models, chosen frames of reference, or even specialization. There is also an emotional, psychological level of expectations, values, and needs. In perception we form a system with what we observe which involves input and its encapsulations. But *patterning* of sensations involves preconceptual models as well as input. Perceptions are "coded in terms of expectations formed from previous perceptions." The sociocultural level includes categories of language, experience, and professional paradigms and practices. And by its very nature, selectivity, at best, can offer only partial pictures. That is, the creative yet limited character of human physiological, intellectual, imaginative, and even emotional equipment, as well as our accumulative though incomplete technical equipment and scientific knowledge, give us little more than partial pictures to augment and to rework at any time. The fable of the blind men and the elephant well illustrates the point of our projecting partial pictures upon nature. We too often identify these partial pictures as full, even final mirror images of nature rather than incomplete impressions based upon our makeup and surroundings. The error lies in our failure to distinguish between our partial pictures of reality and reality itself due to our limited knowledge, makeup, tools, and experience.

Chapter 12, particularly the section on light, sight, and color, has already emphasized the creative but limited character of rainbow formation as an interaction between sunlight, clouds, and us. Let us now review in more detail the selectivity of our senses and minds and their limiting but creative characteristics that form the basis of our partial pictures of nature.

Human consciousness—including awareness that we are aware—is an exquisite entity that seemingly has evolved to serve us in biological survival. Multitudes of ever-changing stimuli bombard us from all directions. Our sense organs and brain save us from being overwhelmed by filtering out stimulation or information irrelevant to our immediate purposes or long-term survival. Regardless of the benefits, such filtering reduces the data from which the brain forms partial pictures.

On the other hand, in a dark forest we may become aware of sounds that would otherwise escape us. Fears may amplify certain sounds against others. That amplification serves survival regardless of other considerations.

The human eye responds to only one out of about sixty octaves of the electromagnetic spectrum. We arbitrarily label that one octave of vision, light waves. Science partly overcame that handicap of selectivity in ordinary human vision through radio (wave) receivers and X-ray tubes and screens. Such technological devices extended human senses to give evidence of a very broad electromagnetic-wave spectrum. These waves outside of light and heat radiation had been from the beginning bombarding us without our awareness. Evidently we hadn't needed information about these unknown waves from the cosmos for our survival.

If one works (or used to work, in this age of computers) in a large room of many pounding typewriters, the ears after a while become accustomed to the constant input so that the pounding recedes almost to the point of unawareness. The brain tends to immunize constant sensory input in favor of new or changed stimuli as when a fire drill siren suddenly sounds in the midst of those dinning typewriters. The human ear, like other sensory organs, evolved to discard extraneous information while retaining input for survival. In any case, the human ear does respond to mechanical vibrations within a range from about 20 to 20,000 cycles per second. However, this has not prevented the brain from developing *ultra* sound concepts, equipment, and techniques to overcome limitations of the human ear. The brain obviously structures, reduces, or extends sensory data.

The simple eye mechanism of the frog affords a case in point. Only the following four kinds of sensory information are forwarded to the comparatively simple brain of the frog: (1) a general bare outline of the frog's immediate environment, (2) sudden moving shadows, (3) sudden changes in light, and (4) the presence of small dark objects nearby through the frog's insect-perceiving mechanism. Just about all other information is excluded from the frog's vision. And yet, so far, frogs have survived; natural selection did not seem to require more acute vision. In our much more complex human visual faculties we have specialized cells in the retina and visual cortex that detect input *changes* without being responsive to constancies.

To further understand the knowledge process in terms of interaction, let us generalize as follows:

1. Knowledge is structured in the human mind by what we bring to data individually and socially. Witness, for example, the variety of responses in the well-known Rorshach inkblot tests.

2. In biological survival the brain selects and structures already filtered sensory data. For example, the brain reverses the upside-down sensory image of the outside world on the retina.

3. Our awareness of an external reality reflects our internal selective equipment. That is, we deal with our internal images as shaped by data, selective senses, brain, and previous individual and social experiences.

4. Encapsulations, as we shall see, exist in our language faculties as well as in limitations of our personal and social experiences.

5. Encapsulation also exists in our being an active part of the interactive knowledge process. Can we be part of the process and still be detached or fully objective? This was discussed in an earlier chapter.

6. If human awareness marvelously emerged from nature, to what extent does order emerge from randomness or chance as we, in scientific and other cultures, attempt to call the world to order?

7. Man is both a subject and an object. The subject studies himself as an object in a world of objects.

Two Brain Hemispheres

The brain is divided into two hemispheres. Each in general has its own specialization and yet is interconnected by a bundle of fibers called the *corpus callosum* to the other hemisphere. Thus in the brain's makeup we discern polarity, interconnection, and integration. And yet this can be encapsulating as well as creative in accordance with individual personality tendencies, abilities, education, and use.

First, we'll discuss the specializations with their considerable division of labor. The left brain hemisphere, connected to the body's right side, is primarily involved in analytical, logical thinking as in mathematical and verbal processes. These processes are step-by-step and linear, and in this sense involve time. The mode is rational and verbal. The right hemisphere, which connects to the body's right side, may be said to be holistic,[3] more spontaneous. Images, answers, or processes arrive as in the intuitive flash whether it be in artistic vision, dancing, recognition of a person, or possibly a hypothesis formation. The sudden holistic aspects here are more space-like than time-like. The hypothesis formation and hypothesis testing of Newton and Einstein, discussed in previous chapters, illustrates the creative integration in science of right-brain intuitive and left-brain analytical modes.

Today we witness the phenomenon of the brain investigating itself, which should serve to extend its creativity and its limitations, and perhaps even transcend former barriers in dealing with itself and its surroundings. For example, let us consider research on the two brain hemispheres. Robert E. Ornstein points out that as far back as 1864 a noted neurologist, Hughlings Jackson, cited "a patient with a tumor in the *right* hemisphere who no longer recognized object, persons or places" (italics added).[4] Certainly this provides early evidence of a relationship between the right section of the brain and holistic, spatial recognitions.

On the other hand, Ornstein continues, "In hundreds of clinical cases, it has been found that damage to the left hemisphere very often interferes with, and can in some cases completely destroy, language ability. An injury to the right hemisphere may not interfere with language performance at all, but may cause severe disturbance in spatial awareness, in musical ability, in recognition of other people, or in awareness of one's own body. Some patients with right-hemisphere damage cannot dress themselves adequately, although their speech and reason remain unimpaired."[5]

In well-known "split-brain" experiments done in the mid-twentieth century, Professor Roger Sperry of the California Institute of Technology "disconnected" the two halves of the brain by severing the corpus callosum of laboratory animals,[6] with results that led to quite successful application to human patients with severe epilepsy by Dr. Joseph Bogen (an associate of Dr. Sperry) and Dr. Vogel of the California College of Medicine.[7] The idea was that with seizure in one brain hemisphere, severing the corpus callosum would result in the other hemisphere taking over body control. In many instances, the extreme severity of the epilepsy attacks were reduced enough for patients to return home with very little abnormality. But through experiments well-suited to the purpose, Sperry's team found that the radical operation helped to establish the specialization of the brain hemispheres. When, for example, a patient fingered a hidden object, say a pencil, in his right hand, he could correctly describe it. If a hidden object was in his left hand, the patient could not describe it. Since the right hand is connected to the left verbal brain hemisphere, the "split brain" patient could adequately describe the unseen pencil or other object with that hand. The left hand is connected to the right "nonverbal" hemisphere; therefore, the patient could not describe the unseen object by using that hand. Thus we possess creative and yet limiting sensory, intellectual, and intuitive apparatus.

In further experiments with split-brain patients, Dr. Bogen tested their ability to write and draw with each hand. Supposedly, writing involves the verbal left hemisphere; and drawing the more spatial right hemisphere. After surgery, the right hand of right-handed patients retained the ability to write English (verbal left hemisphere), but in considerable measure lost the ability to draw. The relational, spatial contribution of the right hemisphere through the corpus callosum was not there. The opposite was true when the left hand of these patients was tested. The left hand (spatial right hemisphere) could still draw and copy spatial, e.g., geometric, figures, but was not able even to copy written words; it could perform graphically, but showed almost no ability to transmit analytical, verbal information. Reason and intuition complete each other creatively. Reasoning, with its parts analysis, seems to involve the systematic, step-by-step conscious mind (left brain hemisphere). Intuition in its holistic syntheses and insights, among other possibilities, may draw upon the Freudian "unconscious" with its psychobiological mechanisms and "dreams," by way of the left brain hemisphere.

Daniel Goleman, associate editor of *Psychology Today,* reported in the May 1976 issue the amazing near-recovery of a Portuguese girl who received a very severe blow on the right side of her head when she was five years old. She was left half-paralyzed and could see poorly and only on the left side of her visual field. At age nine, epileptic seizures began to take over until by age twenty, the seizures grew so severe that drugs were of no avail. By this time, rage and aggression had become characteristic of the young woman. In desperation, the doctors surgically removed the entire right side of the young woman's brain. Within a month after this drastic operation, the seizures and paralysis completely disappeared and the patient could walk. Fourteen years later, the patient was employed doing housework and taking care of a child. She needed no drugs and her behavior was normal.

The results were amazing and difficult to explain: The right hemisphere was gone, but the left side of the woman's body did not show any expected serious handicaps. Also, her right eye did not suffer a loss of vision in the left side of its visual field as before the operation. This had been expected since the inside half of each eye is connected to the opposite brain hemisphere, and the outside half to the same hemisphere. Indeed, the right eye's visual field had greatly expanded although the left eye remained nearly blind. There were no problems with spatial organization, movement coordination, or recognizing music and other nonverbal sounds, all such abilities generally associated with the right brain hemisphere. According to Dr. Coleman, Dr. E. Roy Brown (the research team head of New York Medical College's Brain Research Laboratory and inventor of quantitative electrophysiological battery tests) contends that in a case such as the above, "the remaining brain has learned to use other cells for the functions previously controlled by the removed brain section."[8]

In short, our brain, nervous system, and senses are structured to process external and internal stimuli in particular, selective ways and to compensate in some losses of function. Recognition of the structuring and selective processing enables realization of further creative possibilities within limitations. Thus, in terms of "objectivity," how we perceive reality depends upon our perceptual and physical makeup, our mental and organizational equipment, and our technical advances as well as upon our linguistic, professional, and sociocultural advances and encapsulations.

ENCAPSULATION

> For centuries we have been conditioned by nationality, caste, class, tradition, religion, language, education, literature, art, custom, convention, propaganda of all kinds, economic pressure, the food we eat, the climate we live in, our family, our friends, our experiences—every influence you can think of—and therefore our responses to every problem are conditioned.[9]

Webster's New Twentieth-Century Dictionary (1983) defines encapsulation as "the act of enclosing in a capsule." Psychology professor Joseph R. Royce, in his *Encapsulated Man*, extends the definition conceptually to "claiming to have all the truth when [at best] one only has part of it."[10] That is, if in science, philosophy, religion, society, psychology, music, art, language, or just plain living, we think that we have the final, absolute answer to anything, we are encapsulated. Nature, life, knowledge seem to seep through any trap that we set for it intellectually, intuitively, or imaginatively. Nature and knowledge are expansive and exist without walls in a full sense. There always is a background frame of reference.[11] Besides, we are comparatively too ignorant and lack much too much physiologically, psychologically, intuitively, and otherwise for any final pictures. Also, we are interactors, not just observers. Yes, the blind men and the elephant again!

Encapsulation shows itself in limitations in our language, definitions, concepts, theories, and conceptual models as well as in background frames of reference. It again appears in our personal, cultural limitations as well as in the scientist's role as interactor in nature.

Cultural themes, for example, can be creative and yet limited. Take the cultural theme of the rise, development, and decline of institutions, ideas, and practices, perhaps based upon our individual life pattern. Oswald Spengler, in his *Decline of the West* (1928),[12] contends that most civilizations pass through predictable cycles of rise, development, and decline with implications that current Western civilization has reached a materialistic stage of decline. The rise and fall of ancient Babylonia, Egypt, Greece, and Rome are cases in point of former cycles. The nineteenth-century German philosopher Georg Hegel, in his *Philosophy of History,*[13] emphasized an idealist dialectical principle of change in a universal process, in which the interacting polarities of *thesis* and *antithesis* are transformed into a *synthesis.* Perhaps Hegel took a cue from the ancient Greek natural philosopher Heraclitus (sixth century B.C.E.), who ascribed change in the natural flux of things to "strife" between opposites in nature.

The German social philosopher and revolutionary Karl Marx materialistically applied the Hegelian dialectic of thesis, antithesis, and synthesis to what he considered to be the rise, development, and decline of contemporary capitalist society in his *Das Kapital.*[14] To Marx, the capitalist means of production, to fuel its fires, must expand its markets internally and externally against entrepreneurs of other nations with the same objectives. Thereon hangs a tale of modern economic crises, unemployment, wars, and unequal social conditions and opportunities.

Then comes the theme of rise and fall in the realm of physics. Einstein and Infield in their book, *The Evolution of Physics,*[15] title their first three sections: (1) The Rise of the Mechanical View; (2) The Decline of the Mechanical View; and (3) Field, Relativity. Einstein and Infield contend, in some detail, that assumptions and other encapsulations in the rise, development, and decline of the formerly highly successful Newtonian mechanics led, through eventual anomalies, to Einstein's theory for solution and absorption: Newtonian physics became a special case of Einstein's theories of relativity.

More recently, Thomas Kuhn's *Structure of Scientific Revolutions*[16] has emphasized the rise and fall of what he refers to as "paradigms" in science that dominate conceptual models and associated scientific practice at any given time. These dominant or "normal" paradigms are eventually eclipsed or absorbed by contending paradigms because of accumulated anomalies arising from the original paradigms.

Let us take ancient Chinese science as a type of cultural encapsulation. The ancient Chinese were empirically inventive. According to Dr. Joseph Needham,[17] recognized as the world's foremost authority on ancient Chinese science and civilization, they invented the iron plow with moldboard as well as the collar and trace harnesses for farm animals—great boons for early agriculture. They were the first to discover paper and use paper money, umbrellas, wheelbarrows, fishing reels, multistage rockets, guns and gun powder, magnetic compasses, hot-air balloons, brandy, the chess game, parachutes, porcelain, silk, the printing press, silk cloths, the steam engine, the suspension bridge, and a civil service system. On the other hand, the Chinese practiced a broad Taoist natural philosophy of *yin* and *yang,* polar opposites designated as female and male, that in affinity universally formed various entities of nature. Taoist philosophy portrayed nature as flowing, dynamic, and ever-changing with all things holistically interrelated. Yin and yang supposedly created a dynamic tension throughout nature that effectively drove the changing character of life and things.

Until the Industrial Revolution, Chinese science was in advance of Western science for many centuries. In the interest of this chapter we ask: Why did China not maintain its lead in science with a follow-through Industrial Revolution of its own, which is only now being pursued there in the late twentieth century? Briefly, China fell behind for political and economic reasons. A highly centralized and organized feudal bureaucracy did not promote scientific investigation as such. The successful rise of capitalism from mercantilism in the West overcame feudalism there and stimulated scientific research and development. A marriage resulted between theories of nature and everyday application. In China, there remained a dichotomous gap between natural philosophy and empirical inventiveness; testable hypotheses were not developed to relate theory and practice.That is, the ancient Chinese in their natural philosophy saw the forest and not the trees; as yet they lack Western experience in scientific analysis and application. We Westerners, on the other hand, see the trees but seldom the forest. Encapsulation exists in both cases. Mature science emphasizes both the forest and the trees.

Thus, cultural threads seem to run historically through ideas and mechanisms of change to expand knowledge and yet, at any time, to direct and limit its expansion. In our comparative state of ignorance of nature and of human limitations, this all seems like an inevitability of perpetual revolutions in scientific achievement. Yet today's tragic lag of social intelligence behind scientific and technological advances has placed civilization in terrible jeopardy.

Let us further summarize aspects of encapsulation in the creation of scientific knowledge:

1. *Knowledge is relative to human senses and sensibilities.* The fable of the blind men and the elephant has been cited a number of times. Another example already given is that light covers only one octave in the 60-octave electromagnetic spectrum. For ages human beings were aware of only that one octave because of their sensory limitations. Yet even today, how many unknown signals (knowledge data) from the cosmos are bombarding us because we do not have the physiological, conceptual, or other equipment to track them? It is what we know—and also what we do *not* know—that shapes our observations and explanations.

Another example of the physiological encapsulations of knowledge is this: A motion picture, run at up to sixteen frames a second, is seen only as a succession of individual frames. At that point and above, our visual apparatus fuses the individual frame impressions to give us life-like movement. In this case, the physiological limitation of frame images at the point of fusion is beneficial: we have a choice of individual frame images or movies.

What any organism brings to its surroundings determines its limitations or encapsulations. Whether for better or worse depends upon circumstances. We have previously cited the four kinds of sensory information that limit a frog's vision due to its very simple eye mechanism. The paramecium, a one-celled organism, has only one response, flight reaction, to all outside stimuli, whether chemical, thermal, tactile, or other.

2. *Knowledge is relative to tools.* Do light rays travel as waves or particles? When light rays are sent through prisms (Newton[18]) or diffraction gratings, light shows wave characteristics. When prisms or diffraction gratings were replaced by photoelectric cells (Einstein[19]), light shows particle characteristics, that is, are explainable as bundles of energy called photons. But how can light be both a continuous wave and a discontinuous particle? Perhaps waves and particles are aspects of a bigger picture. At different angles "the neck of the peacock shows different colors" (Seneca[20]). Even though we are encapsulated, the scope and precision of future apparatus and new tools—and new concepts to which they may lead—are a long way from being exhausted.

3. *Knowledge is relative to operations selected.* We saw this in the case of the nature of light above. But, on the other hand, operations are selected by experience. That is, operations lead to new knowledge, and new knowledge in turn leads to new operations. For example, concepts of unseen radio waves longer in wave length than light on Maxwell's electromagnetic wave spectrum (as well as optical telescope development) eventually led to experiments (operations) in the development of radio telescopes. Knowledge of *natural radioac-*

tivity, as in radium (Madame Curie[21]), led to successful operations in man-made artificial radioactivity, including nuclear chain reactions and nuclear reactors (Enrico Fermi).[22] That is, some atomic nuclei break down naturally, resulting in the disappearance of some matter and the creation of energy. In *artificial* radioactivity, mankind through scientific reasoning and apparatus artificially breaks down atoms to release tremendous energies.

4. *Knowledge is relative to mental tools.* Ideas are powerful tools, too. Definitions, hypotheses, theories, and conceptual models are relative to their times and merely approach reality, said John Dewey in *How We Think* (1910). The Ptolemaic-Copernican issue as described in earlier chapters provides pertinent examples. We will talk more about Dewey's *instrumentalism* theory in chapter 24.

5. *Concepts are relative to given areas.* Heat and temperature are often explained in terms of energy of motion of molecules (kinetic-molecular theory). Electrons are subparticles of atoms and molecules. Therefore, since electrons as components are at a level below that of molecules, the concepts of heat or temperature cannot be applied to electrons in terms of moving molecules. The moving-molecule concept is encapsulating when applied to areas below that of the molecule. Another example of encapsulation due to concepts being relative to given areas is furnished by the differences in astronomy between the units of *centimeter, angstrom,* and *light years.* The centimeter or inch is an appropriate unit for small distances on the earth's surface. But since an angstrom is only one hundred-millionth of a centimeter (1 cm = .3937 inches), angstroms are used to measure the length of light waves which are much too minute for use of the centimeter or inch as a unit of measurement. On the other hand, distances are so vast in cosmic space that even the mile or kilometer is inappropriate. It takes light, traveling more than 186,000 miles per *second,* over four years to arrive on earth from the nearest star, Arcturus. Very distant galaxies send light to us that has traveled *billions* of years before reaching us. Obviously, a new unit of distance was needed for such vastness. That unit became the *light year,* which is the distance that light travels in one year, or six trillion miles. That means the *nearest* star is about 25 million miles, i.e., over four light years, away. For comprehension, different units are needed for the minute, the everyday, and the macroscopic in handling distances in the physical world.

6. *A problem of extrapolation exists in trying to apply former concepts to new areas.* Sudden leaps often occur in the form of leads during the attempted extension of established horizons. Galileo centered his attention on gravity as acceleration of falling bodies on the earth's surface, and even briefly alluded to a gravity force behind such acceleration. Newton extrapolated the gravity concept to the entire universe of objects, huge and minute, in his universal law

of gravitation with the aid of his laws of motion. This extrapolation, tremendously successful for two centuries, established classical mechanics and astrophysics. The genius of Newton's creativity was and is apparent. But the encapsulation of one of the world's greatest masterworks in science, the *Principia Mathematica* (1687) (see chapter 9), lies in its being temporary, a stage to further knowledge. The encapsulation in this case became dramatically apparent, at least in hindsight, when in 1899, A. A. Michelson, greatly renowned for his experimentation on the speed of light, proclaimed, with typical Newtonian confidence, at a top physics conference at an Eastern university, that "the important fundamental laws and facts of physical science have all been discovered, and these are now so firmly established that the possibility of [their] ever being supplanted in consequence of new discoveries is exceedingly remote. . . . Our future discoveries must be looked for in the sixth place of decimals."[23] Scientific progress seemed a matter of improved techniques, of greater precision of measurement. Scope, structure, and standards of operation seemed fully achieved.

Overlooked in Michelson's Newtonian encapsulation was the irony that the atom was found, also at the turn of the twentieth century, to have electrical components termed protons (positive) and electrons (negative). Atoms of a given chemical element, for example, radium, were transmuting into another element, radon. In their natural radioactivity, atoms emitted radiations termed alpha, beta, and gamma rays. All this, among other anomalies, contradicted Newtonian physics based on indestructible mechanical atoms without electrical or other components.

Consequently, with the appearance of Einstein's special theory of relativity in 1905, the props of the formerly firm mechanics were shaken, holding only under limited conditions. Newton's mechanical model, so ingenious in the intermediate world of everyday experience, did not adequately cover the submicroscopic and macroscopic realms. But only future scientific development could reveal that.

Another example of the temporarily successful application of a model from one area of physical science to another involves the great nuclear physicist Lord Ernest Rutherford (1871–1937), who, on the basis of his very famous gold foil experiments, concluded in 1911 that atoms are like miniature solar systems. At the center of each atom, he claimed, is a tiny, highly concentrated nucleus of positive charges. Around the nucleus, electrons revolve in different orbits like planets. Between nucleus and electrons are vast reaches of space such as comparatively exist between sun and planets. This application of a planetary model to minute atomic structures worked well for the simplest of atoms, hydrogen, but failed for the more complex atoms above hydrogen.

This planetary model, however, led to a highly innovative atomic model by Denmark's great physicist Niels Bohr (1885–1962). The Bohr model was a first in applying early stages of quantum theory to atomic electron orbits and properties. Electrons jump between allowed orbits emitting such electromagnetic radiation as X-rays or light rays. This successful idea in turn led to current quantum-mechanical models. As we can see, then, extrapolation of former concepts to new areas may not always be successful; however, safeguards against extensions or applications of concepts to new scientific areas bring to mind Francis Bacon's admonition in the seventeenth century that scientists must become factually or empirically familiar with the new areas.

7. *Every thought and action involves basic assumptions and is encapsulated by these assumptions.* Euclidean geometry was a productive encapsulation of Newtonian physics. Einstein's relativity broke through that encapsulation by being based upon non-Euclidean geometry and its assumptions.

8. *Language sharpens thought and provides shades of meaning.* Man's language, like his tool-making abilities, has given him great survival and cultural advantages that in time can be very encapsulating, however. More specifically, words are like labels for identification of things and ideas at any stage of knowledge. But things and ideas change and so do our experiences and knowledge of them. Percy Bridgman[24] has emphasized *operational definitions* for science, whereby definitions change in accordance with our experiences with changing things, ideas, processes, and institutions. For example, the final definitions of matter, energy, or even democracy in their ultimate scientific, economic, political, and social aspects are certainly not yet in sight.

Mathematics, a language and logic of science, is definitely precise, concise, and consistent with a minimum of emotional and semantic overtones as compared to words as symbols. But as previously emphasized, conclusions based upon mathematics are no more extensive or complete than the basic assumptions of any given mathematical system. Witness again the revolutionary results that Einstein obtained with non-Euclidean mathematical assumptions in his relativity theories. Time became a fourth dimension of space, and the physical universe was made understandable in terms of a four-dimensional space-time gravitational field—a long way from Newtonian universal gravitational forces in an assumed three-dimensional Euclidean world. With Einstein a straight line was no longer necessarily the shortest distance between two points, as on the curved surface of the earth or in a curved space-time. What will replace Einstein's relativity in the future to make it a special case of a still larger picture?

Meanwhile, we remain with challenging obstacles to true objectivity (subjectivity, selectivity, and encapsulation), as well as with personal human contact points with nature: senses, muscles, mind, imagination. But we also have at our disposal some technology as extensions of ourselves and some background interaction experiences with nature.

NOTES

1. Alfred North Whitehead, *Science and the Modern World* (New York: Mentor, 1958).
2. Albert Einstein and Leopold Infeld, *Evolution of Physics* (New York: Simon and Schuster, 1952).
3. Although I am using the usual *holistic* spelling, I would have preferred *wholistic*.
4. Robert E. Ornstein, *Psychology of Consciousness* (San Francisco: Freeman, 1972), p. 69.
5. Ibid., p. 71.
6. Ibid.
7. Ibid., pp. 71–76.
8. Ibid., p. 75.
9. J. Krishnamurti, *Freedom from the Known* (New York: Harper & Row, 1969), p. 25.
10. Joseph R. Royce, *The Encapsulated Man* (Princeton, N.J.: Van Nostrand, 1964), p. 2.
11. We can even claim that the universe as a whole may be expanding into a vacuum background.
12. Oswald Spengler, *Decline of the West* (Oxford: Oxford University Press, 1928).
13. Georg Hegel, *Philosophy of History* (Colonial Press, 1900).
14. Karl Marx, *Das Kapital* (New York: Modern Library, 1936).
15. Einstein and Infeld, *Evolution of Physics.*
16. Thomas Kuhn, *Structure of Scientific Revolutions* (Chicago: University of Chicago Press, 1970).
17. Joseph Needham, *Science and Civilization in China,* vols. I–V (Cambridge Viking Press, 1960–). Eight volumes are planned and still in progress. The Chinese History of Science Society in Beijing considers Joseph Needham and his associates to be the outstanding researchers of ancient Chinese science and society.
18. Isaac Newton, *Opticks* (New York: Dover, 1952; based on the 4th ed., London, 1730).
19. James Perlman, *The Atom and the Universe* (Belmont, Calif.: Wadsworth Publishing Co., 1970), p. 388–90.
20. Lucius Annaeus Seneca, *Naturales Quaestiones* (London: Macmillan, 1910).
21. Eve Curie, *Madame Curie* (Garden City, N.Y.: Garden City Pub., 1943).
22. Perlman, *The Atom and the Universe,* pp. 535–43.
23. Ibid., p. 470.
24. Percy Bridgman, *Logic of Modern Physics* (New York: Macmillan, 1927).

SUGGESTIONS FOR FURTHER READING

Asimov, Isaac. *The Human Brain.* Bergenfield, N.J.: New American Library, 1965.
Blakemore, Colin. *Mechanics of the Mind.* New York: Cambridge University Press, 1979.
Borek, Ernest. *The Atoms within Us.* New York: Columbia University Press, 1967.
Calden, Nigel. *The Mind of Man.* New York: Viking, 1973.
Campbell, Robert. *Enigma of the Mind.* New York: Time-Life Books, 1976.
Gregory, R. L. *Eye and Brain.* Cambridge, England: World University Library, 1965.
Hooper, J., and D. Toresi. *The Three-Pound Universe.* Los Angeles: Tarcher, 1986.
Love, David. *The Sphinx and the Rainbow.* New York: Bantam Books, 1984.
Oatley, Keith. *Brain Mechanisms and the Mind.* New York: E. P. Dutton, 1972.
Ornstein, Robert, and David Sobel. *The Healing Brain.* New York: Simon & Schuster, 1988.
Ornstein, Robert, and Richard Thompson. *The Amazing Brain.* Boston: Houghton Mifflin, 1964.

Royce, Joseph R. *The Encapsulated Man.* Princeton, N.J.: Van Nostrand, 1964.
Scientific American Special Issue. "Mind and Brain," September 1991.
Scientific American Readings. *Progress in Psychobiology.* San Francisco: W. H. Freeman, 1976.
Wilson, John R. *The Mind.* New York: Life Science Library, 1964.

14

Expanding the Cocoon

> Basic problem-solving steps are the following: felt difficulty, problem clar-
> ification, suggestions for solutions, suggestion selection, and verification.
> John Dewey, *Experience and Nature,* 1923[1]

Dewey's "steps" for scientific problem solving, as given above, run in a formal, consecutive order. The present writer prefers the following more flexible
modification: basic problem-solving *aspects in various orders and combinations* instead of conceptual formal steps. Such more flexible problem-solving
aspects vary with specific observational and experimental situations, as
demonstrated in chapters 1, 2, 4, 6, and 9–11, particularly chapters 1, 2, 10, and
11. This will be further amplified in the following chapter by tracing the
expanding partial pictures of the rainbow from Seneca to Roger Bacon to
Newton. Meanwhile, let us first consider the following abilities, attitudes,
and considerations that have expanded cocoons of interpretation.

CONSIDERATIONS FOR OBSERVING AND CLARIFYING

1. *We cannot trust our senses or appearances alone.* For example, the earth
seems to us stationary, with the sun, moon, planets, and stars revolving around
us (chapters 4–6, the Ptolemaic-Copernican issue).

2. *There are both similarities and differences in things.* Analogies based on
similarities have afforded fruitful leads in solving problems. But the difficulties, sometimes dangers, of analogies lie in not knowing the differences
between things compared. The Copernican solar system led to the Rutherford-
Bohr analogy of the planetary model of the atom, which enjoyed early success:
The planetary atom accurately predicted radiation from heated or electrically

excited hydrogen gas; it explained the origins of light, X-rays, infrared and ultraviolet radiation. Even now it explains the origin of radio signals from hydrogen in the stars. It accounts for magnetism, and it has served as an effective model to explain chemical behavior as well as regularities in the periodic table. And yet when the model was used to predict the spectral colors and radiation frequencies from elements heavier than hydrogen, it got into difficulties. Predictions for helium were fairly accurate, but the higher the element the lower the accuracy. Replacing circular with elliptical orbits worked well for Kepler's planets, but not adequately enough for Bohr's electrons when the German physicist Arnold Sommerfeld (1868–1951) tried it. Bohr's energy levels and electron distributions still serve many purposes, but the concept of an "electron cloud," representing the probability of an electron's being in a particular place at a particular time, has replaced Bohr's idea of set electron orbits.

To look for exceptions or contradictions to findings, analogies, and ideas can be a safeguard against encapsulations. For example, Tycho Brahe, at the end of the sixteenth century, was openminded in respect to the Ptolemaic issue. Ptolemaic astronomers claimed that if the earth was just another planet, stellar parallax would have been observed as it revolved around the sun (see chapter 7). But no astronomer had yet observed stellar parallax. Brahe possessed the most precise astronomical instruments to that time. With utmost patience he looked for the undiscovered stellar parallax—but in vain. He therefore remained a Ptolemaic, but on a modified basis. By picturing all the then known planets revolving around the sun as the sun itself revolved around the earth, he moved up to but stopped short of Copernican theory. When Brahe's assistant, Kepler, inherited Brahe's planetary data, he found that the data could show that the earth as well as the known planets moved around the sun if orbits were *elliptical*. Thus, Brahe's unsuccessful search for stellar parallax resulted in Kepler's elliptical rather than circular Copernican orbits of the earth and other planets around the sun.

Science has been very successful in emphasizing polarities in nature, as north and south magnetic poles, positive/negative and electrical charges, and attraction and repulsion effects. Further, antimatter was hypothesized as polarity to ordinary matter. With protons established as a building block of matter, Paul Dirac daringly predicted in 1930 the existence of antiprotons in line with his antimatter postulate.[2] Emilio Segre artificially created antiprotons in 1955 at Berkeley, California.[3] Positrons, positive electrons, had previously been found by Carl D. Anderson in 1932 as predicted by Dirac. Einstein, to further his unified field theory, attempted to work out a theoretical basis for predicting the existence of an antigravity force (levity), but was unsuccessful. Searches for evidence of an anti-universe are currently on the agenda. There are benefits in a guiding principle of approaching nature in terms of polarities. It has expanded encapsulations, notwithstanding Einstein's failed attempts at antigravity.

3. *Definitions also play a role in clarification of ideas.* Operational definitions in which concepts are defined in terms of the actual processes in which they are experienced was discussed in chapter 1 as working definitions. Light first became understandable in terms of electromagnetic waves when associated with lens, prism, or diffraction-grating experiments, but as photon bundles of energy when photoelectric cells were used. Waves have continuity; bundles of energy are discrete.

Aristotelian definitions also have a place in clarification. They are based on classifying objects or ideas according to observed similarities but with indication of the individual difference of a given object or idea within the expressed classification, much as any of us is characterized by both a family and an individual first name.

4. *The use of authorities involves careful considerations.* Authority is relative to given fields. Would a civil engineer necessarily be more knowledgeable in general nutrition than you or I? Even in the same field, equally able and experienced authorities do not always agree as, for example, in medical opinions on psychosomatic medicine.

Individual thought, however, with first-hand evidence in addition to use of authority, is necessary for advances in any scientific area. Successes of scientific research generally attest to that. The best basis for judgment of a source's value is the training, purpose, and experience of the author. For example, can the opinions of a physician working for a tobacco company be as objective and dependable with respect to nicotine's dangers to health as those of an independent biochemist?

5. *Variation, change, and motion are common to all things.* Whether we consider our own individual lives, the history of any nation or idea, or our expanding universe, we must take into account both changes and variations (see chapters 1 and 4). Historically tracing the expansion of explanations of the rainbow in the next chapter will describe basic factors in the enlargement of ideas with time.

6. *Look for the basic elements, factors, and variables in a problem situation.* This is what Francis Bacon meant when he facetiously exclaimed, "Go to the horse's mouth to count its teeth." We have seen that Brahe had the right idea when, in trying to decide whether to be a Ptolemaic or a Copernican, he spent much time carefully plotting the planetary path of Venus against the background of stars (chapter 7). Today, geology has developed a broad theoretical base of its own in its continental drift theory, but still cannot adequately predict earthquakes. More detailed knowledge of factors and variables of earthquakes are needed. Meteorology as a science has a record of only about 80 percent success in weather prediction, even after a vast accumulation of knowledge about

weather and climate in the twentieth century. More knowledge of factors and variables is needed in specifics of air temperature, pressure, humidity, cloudiness, and rainfall at given times and across time. Of course, conceptual models have already emerged from specifics, such as traveling *highs* and *lows,* cyclones and anticyclones, relative humidity, cloud classification, water cycles, weather fronts, and convectional air circulation or jet streams. Increased knowledge, for example, about basic elements, factors, and variables in upper regions of the atmosphere, such as the stratosphere and ionosphere, could raise successful weather prediction above present probabilities.

7. *Know the difference between a fact, a theory, and a definition.* A fact is an event, an actuality. A theory is a speculation, an explanation. In science, a theory is often an underlying explanation of what may be observed or experienced and is to be differentiated from what is immediately observed or experienced. For example, is an atom a theory or fact? To the ancient Greek natural philosophers, it was a theory. To them, the atom, by definition, was the smallest particle of a given substance. Speculatively, all matter in the universe "consisted of atoms and the void."[4] The atoms were in constant motion and varied in size and shape according to substances. Rapidly moving atoms were a speculative idea that was used to explain many physical phenomena, such as wind, water currents, melting, and freezing. With successful application, the speculation advanced in status from hypothesis to theory. At the beginning of the twentieth century the renowned physicist-philosopher Ernst Mach still considered the atom to be a convenient fiction. It wasn't until the invention of the electron microscope later in the century that atoms were photographed as actual units within molecules. With such visualization, atoms became facts.

An example of knowledge or truth by definition rather than by fact is the statement that 10 millimeters equal 1 centimeter. That is, if a space interval arbitrarily set off as a centimeter is divided into 10 equal parts, then *by definition,* 10 mm = 1 cm. Or, when the freezing point of water is arbitrarily set at 0° Centigrade and the boiling point of water at 100° Centigrade, by definition a Centigrade degree equals $\frac{1}{100}$ of the temperature interval between the freezing 0° C and the boiling temperature of water.

CONSIDERATIONS FOR CLASSIFYING AND GENERALIZING

8. *Classification is based on the grouping of similarities observed among objects.* Differences are recognized within, as well as between, groups. All sciences classify (as do other areas of culture). In biology, an example is taxonomy with its grouping of plants and animals into phyla, species, etc. In physics and chemistry, the periodic table of elements is indispensable, including its

families of elements with differences within a family. In astronomy, classification of solar system data as sun, planets, and moons, or of star characteristics, is basic to similarities and differences. In geology, it is rock or mineral classification. In meteorology, cloud classification is important to weather description and prediction.

9. *Graphs, charts, tables, and co-ordinate systems afford valuable mathematical tools for organization and interpretation of data.* The first three mathematical devices are so general, concise, and visual, including media use, that they really do not need to be illustrated here. Latitude and longitude to mark the earth's surface, for example, and their equivalents on the celestial sphere point up the indispensability of charts, graphs, and tables for object or area location as well as for general space organization.

10. *Tentative, careful guessing can be very fruitful and productive in problem solving.* But generalization that is either too hasty or too cautious blocks progress. Being open-minded, systematic, and critical can help here, as can knowledge and experiences in a given area. But remember, "It is easier to be critical than to be correct," said Benjamin Disraeli.

11. *Modern advances in technology and medicine reflect tremendous benefits from looking for causes of phenomena in the workings of nature itself.* That is, searching for cause-and-effect relationships in nature has been a powerful approach in classification and generalization.

12. *Other things being the same, the fewer the assumptions in an explanation, the greater the probability of success for that explanation.* This illustrates the principle of parsimony: fewer assumptions mean fewer possibilities of being off the track.

CONSIDERATIONS FOR FORMING CONCLUSIONS

13. *Everyone is entitled to his opinion, but not all opinions can stand up equally under the facts.*

14. *It is more important to be able to anticipate, detect, and correct errors in problem solving than to expect perfect solutions.*

15. *Many leads and hypotheses often have to be tested before the best solution is found.* Conclusions and statements, therefore, should be qualified according to the limits of the particular problem, conditions, and evidence.

16. *Since facts are never completely all in, certainty can merely be approached, not arrived at.* Conclusions, then, are merely the best evidence at

the time, and require openmindedness for further verification, improvement, or change.

17. *In some cases there is more than one correct answer to a problem due to two answers being different aspects of the same thing, or due to the need for additional knowledge.* Witness again the phenomenon of light, which under some circumstances has the characteristics of waves, and under others shows characteristics of discrete particles.

18. *The larger the number of cases as evidence, the longer the conclusion may last in time.* That is, as evidence increases, hypotheses become theories and theories become laws. (See 20 below.)

19. *Every statement, opinion, or idea rests upon some assumption, and is no more solid than its assumption.* The fewer the assumptions the better, as in consideration 12 above.

20. *The hypothesis, the theory, the law, and the axiom indicate degrees of "certainty," actually, of estimates of probability.*

From such generalizations for expanding cocoons of observations and interpretation, let us trace in the following chapter the partial pictures of the rainbow as expanded from Seneca to Roger Bacon to Isaac Newton.

NOTES

1. John Dewey, *Experience and Nature* (New York: Columbia University Press, 1923). John Dewey was an outstanding American pragmatic philosopher, psychologist, and educator who well deserved the distinction he achieved at the University of Chicago and Columbia University.

2. Ivers Peterson, "Strings and Mirrors," *Science News,* February 27, 1993, p. 136.

3. James S. Perlman, *The Atom and the Universe* (Belmont, Calif.: Wadsworth Publishing Co., 1970), p. 557. Segre's artificial creation of an antiproton had been preceded by tell-tale tracks in separate studies by G. Rettalack at Indiana University in 1951, and by B. Rossi at MIT in 1954.

4. Lucretius, *The Nature of the Universe,* trans. Robert Latham (Baltimore: Penguin, 1964).

SUGGESTIONS FOR FURTHER READING

Dewey, John. *How We Think.* New York: Columbia University Press, 1924.
Lucretius. *The Nature of the Universe.* Trans. Robert Latham. Baltimore: Penguin, 1964.

15

Explanation and Interaction:
The Expanding Partial Pictures of
Seneca, Roger Bacon, and Isaac Newton on the Rainbow

Reality is defined by the questions we put to it [at a given time].
John A. Wheeler, *Scientific American,* 1991

In science as in literature, daily life, and various aspects of culture, we deal with partial pictures of what we observe and experience. Development of these partial pictures makes for perpetual revolutions in science. We wish here to trace such changes in ideas of the rainbow from antiquity to the modern era. As investigators of the rainbow we select Lucius Annaeus Seneca (3 B.C.E.–65 C.E.), Roger Bacon (ca. 1214–1294) and Isaac Newton (1642–1727).

That these creative individuals were limited by the accumulated optical knowledge of their times is reflected in their observations, in the conceptual images they form, in their methods for obtaining and presenting information, in their logic for forming conclusions, and in the general character of their problem solving. All this is overviewed in Table 15.1 and will be detailed in succeeding sections. Of course, it will also be seen that each of these observers using the available knowledge of his time, took a creative leap forward.

We shall find that the rainbow also excellently illustrates the development of scientific inquiry from stages of haphazard observation to organized inductive investigation of natural events to controlled laboratory testing.

Most important for our purposes here is the following: The changing, expanding partial pictures that result from each investigator illustrate well how scientists form systems with what they can observe through the use of their minds and imagination based upon the knowledge about premises and practices of their times. At each stage, the differences, specific creativity, and limitations of the *interacting* investigator become apparent as the partial picture expands and changes.

Table 15.1
Overview

	Lucius Annaeus Seneca (3 B.C.–65 A.D.)	Roger Bacon (ca. 1214–1294)	Isaac Newton (1642–1727)
I. Optical Knowledge Premises of the Time	1. Reflection: Geometrics of images formation. 2. No knowledge of refraction or dispersion as concepts or principles. 3. Experimentation not yet a socially developed or socially recognized knowledge tool. 4. Aristotle's *Meteorologica* and its treatment of the rainbow as a distorted reflection of the sun.	1. Geometrical knowledge of reflection. 2. Qualitative geometric knowledge of refraction. 3. Knowledge of optics.	1. Geometrical knowledge of reflection and refraction. 2. Snell's laws of refraction as expounded by Descartes and Archbishop Dominis.
II. Thesis or Hypothesis of Rainbow Maintained or Developed	1. Rainbow as image of the sun distorted by hollow clouds acting as imperfect mirrors.	1. Rainbow as reflected image from clouds of sun's rays, forming the base of a cone of which the apex is at the eye.	1. Raindrops and moisture as refracting, reflecting, and dispersing agents of component colors of sun's rays for rainbow formation.
III. Main Basis for Obtaining Data or Information	1. Memory of own past observations; writings of others.	1. Direct inductive investigation of phenomena.	1. Controlled laboratory setups of considerable inferential character.
IV. Main Bases for Forming Conclusions	1. Reasons from observations, from analogous situations, and from ideas of contemporaries.	1. Inductively generalizes from empirical data of investigation.	1. Verifies original hypotheses by controlled experimentation.
V. Main Method of Presenting the Material of the Rainbow	1. Poses, discusses, and analyses various aspects of the problem.	1. Outlines a series of inductive investigations of various aspects of the rainbow.	1. Systematically tests hypotheses through a hypothetico-deductive system built up from definitions to axioms to the propositions tested. The rainbow emerges as an application of such tested propositions.

SENECA'S THESIS AND THE
KNOWLEDGE BACKGROUND OF THE TIME

Seneca's thesis or hypothesis as revealed in his *Naturales Quaestiones*[1] essentially was that the rainbow is an image of the sun distorted by hollow clouds as imperfect mirrors. In this he is supporting Aristotle's conception in the *Meteorologica*.[2] Seneca, like Aristotle before him, had no conceptions of refraction or its laws with which to work. These men were familiar with refraction phenomena but attributed it to "weaknesses of the eyes." The most advanced ideas of optics of the time relevant to the problem were those of Greek mathematicians involving geometrics of reflection, that is, of mirrors. These, along with numerous speculative writings of other contemporaries, as well as his own observations and ability for brilliant analogous reasoning, were the tools Seneca worked with.

THE ARGUMENT AND THE METHOD

Origin of the Bow

Seneca opens his discussion of the rainbow with the question of its origin. His approach is one of calling attention to *the conditions under which the phenomenon occurs:*

1. Water forced from a burst pipe shows "the appearance of a rainbow."
2. Water "spurted by a fuller" on clothes shows colors of the bow.
3. "A rainbow never occurs except when there are clouds about."

Thus, on the basis of observational evidence, Seneca establishes the cloud and moisture as a first factor in the rainbow's origin. With that he immediately begins his mirror analogy: Each raindrop of the cloud is a mirror of the sun and the "rainbow is a blending of innumerable images." Casting the cloud moisture in the role of mirror, Seneca, in typical proof by analogy, points out that single drops of water on separate leaves will each have an image of the sun. Seneca slips badly, however, when he states as further proof that if a pond is sectioned off each section of the pond will have its own solar image. A little actual checking would certainly have made a difference.

Then, with acknowledgments to Aristotle, Seneca attributes to "weakness of human eyes" the fact that the rainbow is seen as a single blurred mass instead of as many distinct images of the sun. Brilliantly, and without realizing the refraction or bending of light involved, he uses some excellent examples of refraction as evidence of the "weakness of eyes": the broken appearance of an oar in water or the magnified appearance of apples through glass. We believe it important enough to further emphasize the point of the broken oar. Involving the principle of refraction, it really contains a further key to the rainbow, but the accumulated knowledge of the time is as yet not ripe enough for the real significance of the broken oar to be seen in terms of refraction of light. Reasoning and analogy must always be at a disadvantage without sufficient knowledge.

But it has becomes clear that Seneca emphasized the sun as a second factor of rainbow origin. Here his further observations are sharp: "The [rainbow] image is never seen except opposite the sun, high up or low down, *in inverse relation,* just as he sinks or elevates his course" (italics added). Or again: "The bow never appears when the sky is clear, and never when it is so cloudy as to hide the sun."

The third factor of origin, and one that Seneca uses further to emphasize the mirror character of this origin, is the observer, or rather the position of the observer relative to the other two factors, the clouds and the sun: rainbows are never seen unless the observer is between the sun and the clouds. Seneca claims this to be in line with mathematical or geometrical evidence or expectation although he does not offer any details.

In further elaboration of his mirror thesis, Seneca points out that just as objects do not actually exist within a mirror, so does the rainbow not actually exist within the clouds. He again uses observational evidence of sudden appearance and disappearance of the rainbow as when an intervening cloud shuts out the sun.

Colors of the Bow

So much for the origin of the rainbow itself. How did Seneca handle the question of its colors? In seeking an explanation, Seneca looks to what he has established as the basic factors in its origin: the sun and the clouds. Impressed by the fiery color in the rising and setting of the sun and by the "dull" shades of clouds, he attributes the red in the rainbow to the sun and the dark blue to the clouds. He considers the other colors to be an interplay or blending of these basic two. We wish to point out, first of all, the scientifically analytical and empirical character of Seneca's explanation by his going back to the sun and to the clouds for color. It is true that his analysis is merely on a descriptive or qualitative basis, but what other basis could there have been for such analysis nineteen centuries ago? Although he did not have any wave theory of color to work with, notice how very close Seneca is to our theory today of three primary colors of red, blue, and green, with all other colors a blending of these three.

But Seneca is not satisfied merely to explain the variety of rainbow colors as a blending of the primary colors, he is interested in the mechanism involved. He points out that clouds have an infinite variety of shapes and thickness in the transmission of sunlight and that "this difference in consistency causes alternations of light and shade and produces that marvellous variety" of color in the rainbow. Also by pointing out that the angle at which the light strikes is a factor in color, Seneca is getting close to what we consider the relationship between angles of refraction in a prism and color. In further support of his "blending" theory, he uses some analogies that are gems of analogous reasoning, with all their advantages and dangers:

1. Some dyed clothes vary in color according to the angle and the distance at which they are viewed.

2. "A purple garment does not always come out to exactly the same tint from the same dye. Differences depend upon the length of time it has been steeped, the consistency and the amount of moisture in the dye: it may be dipped and boiled more than once."

3. "The neck of the peacock shines with varied colors as often as it is turned hither and thither."

4. A glass stick placed obliquely in the rays of the sun will show colors of the rainbow—in this case, of course, glass taking the place of the cloud.

The first three analogies speak eloquently for themselves. Reflective phenomena are being used as brilliant analogies to support a theory based entirely on reflection. The fourth analogy is of particular interest: It is the correct analogy, based on refraction, but the underlying principle, as with the broken oar, lies hidden, locked beneath. Seneca's prism analogy comes as close to the answer of the rainbow's color as it can without the knowledge of refraction.

Thus, coming as close to the answer as he could for the times, but still falling short, Seneca then moves farther away and even stumbles in his attempt to use this correct analogy as evidence for an incorrect or incomplete theory. To reconcile the fact that in the glass's prismatic color effects, one does not obtain a likeness of the sun, he states that if the glass were symmetrical in shape and "suitably constructed," *it would "reflect as many images of the sun as it had faces.* But since the sides are not distinctly separated from each other, and not bright enough to serve as mirrors, the images get confused through being crowded together and are reduced to the appearance of a single band of color" (italics added). In point of fact, Seneca once more "leaves himself open" by not actually checking the underlying assumption of "as many images as faces." We emphasize again, however, that his is not yet a time when rigorous empirical checking has becomes a criterion of scientific method. It is Roger Bacon, twelve centuries later, who begins to insist on such a criterion. We also see illustrated in the above example the dangers and yet the power that lies in analogies as indispensable tools of science even today, i.e., the dangers of relationships that may be superficial but which yet have the power to reveal underlying truths.

Shape of the Bow

This is Seneca weakest section in the sense that his mirror theory is obviously unable to account for the half-circle nature of the bow:

> 1. Seneca agrees with Posidonius that "the rainbow is formed in a cloud shaped like a hollow round mirror whose form is that of a section through a ball."

> 2. He then uses another authority, Artemidorus, for geometric evidence that such a cloud in the form of a concave mirror would present an inverted reflected image closer to the observer than is the distant object.

So far so good. In 1 above, Seneca is reiterating the cloud-mirror analogy. In 2, he indicates the type of mirror that would give the same relationships between object, image, and observer as actually exist between sun, clouds, and observer, and then infers that rainbow clouds have that shape. The geometric optics involved is correct, however.

Seneca's difficulties begin, however, when he tries to understand why the bow is not a full circle, especially since the halo is a complete circle. All that he is able to do is correctly reject the following two explanations for this half circle and then leave the whole question unanswered:

> 1. The explanation of one "authority" that the cloud can reflect only the upper part of the sun because of the sun's much higher position, he is able to reject by correct observational or empirical evidence that rainbows are often formed with very low setting suns or by the sun in lower positions in the sky than the rainbows formed.

> 2. The explanation of Stoics that the rainbow cloud is a semicircular cloud or semicircular mirror, he refutes by pointing out that the whole of a ball can be seen in a semicircular mirror.

Size of the Bow

Seneca handles the question of the size of the bow, that is, the fact that the bow appears a great deal larger than the sun, not on the basis of the much greater distance of the sun—not much was known in those days about quantitative distances of heavenly phenomena—but by consistently following through with his mirror analogy. He discusses actual mirrors that magnify, stars that appear larger when seen through moisture, and the magnification of objects placed in water. We wish again to emphasize the frequency with which Seneca strives to substantiate his ideas by observational evidence although by analogy. In fact, in the above instance of magnification we particularly note a suggestion to the reader to place a ring in water, a thing which would, from the standpoint of method, constitute an appeal to empirical evidence of an *experimental demonstration* character, This, of course, was an exceptional instance in Seneca's case.

Seasonal Variation

Seneca's brief treatment of seasonal variation of the bow is essentially a restatement of Aristotle's observations which we will merely repeat here:

1. While in the summer, rainbows may occur in the early part or at the end of the day, after the autumn equinox, they may occur throughout the day.

2. The explanation lies in the "greater altitude of the summer sun at midday," and the effects of this in "dispelling clouds" and also in not providing the necessary relative positions between the sun, a rainbow cloud, and the observer.

SUMMARY

We believe that we have adequately revealed the nature of Seneca's problem solving with the rainbow. While it is true that he displays weaknesses—the somewhat rambling and unsystematic presenting of his material, the frequent generalizations from too few cases, and the sloppy evidence—nevertheless, Seneca's handling of the rainbow is science, essentially good descriptive science for his day:

1. He empirically approaches the phenomenon itself.

2. In so doing, he calls attention to the natural conditions under which the rainbow phenomenon occurs, as, for example, the presence of the sun, moisture, an observer, and certain relationships among them.

3. He poses and discusses various aspects of the phenomenon as origin, color effects, shape, size, and seasonal variation.

4. He makes the best use of the accumulated knowledge of his time through brilliant imagination, observation, analogy, and inference.

5. He constantly strives for evidence or "proof," for his arguments by reference to observations, to known analogous situations, to "mathematical

proof" or to "authorities" like Aristotle. If this evidence is mostly but not always empirical, Seneca still emphasizes evidence. And if a great deal of empirical evidence that he does have is of an observational character based on memory rather than of an experimentally analytical or quantitative character, the fault may lie with us in applying today's criterion of rigorous, controlled experimentation to a man who lived nineteen centuries ago. Like everything else, science—its definitions, concepts, methods, and content—changes with time and with the accumulation of knowledge and experience.

In respect to content, Seneca, without knowing about refraction, weaves close to and around it while establishing the origin of the rainbow in an empirical relationship between the sun, clouds, and an observer. If his hypothesis based on a mirror analogy is incomplete or erroneous, he nevertheless makes headway in establishing facts, as with the three basic factors behind the rainbow's origin and their relationships. Does modern physics always do better? In recent years, for example, advances have been made in optics through contradictory and incomplete hypotheses of light consisting of continuous waves on the one hand, and discrete particles on the other.

THE BACKGROUND TO KNOWLEDGE OF OPTICS IN BACON'S TIME

Examination of Roger Bacon's encyclopedic *Opus Majus (The Greater Work)*, particularly his sections on optics,[3] reveals the following general character of his optical knowledge indicated in our overview:

1. He has a well-developed understanding of the anatomy and physiology of the eye, including refractive and optical nerve functions.

2. He has a sound qualitative knowledge of the geometrics of both reflection and refraction. I emphasize *qualitative* since Bacon shows descriptive understanding of general relationships between shapes of mirrors or lenses and resulting sizes and positions of images, but not yet of the precise, analytical, or quantitative basis of a mirror or lens formula as

$$\frac{1}{p} + \frac{1}{q} = \frac{1}{f} \text{ or } \frac{S_o}{S_i} = \frac{p}{q},$$

where p = distance of object to a mirror or lens,
q = distance of image to mirror or lens,
f = focal length of lens
and S_o and S_i = size of object and image, respectively.

Willebrord Snell's law relating the angle of incidence to the angle of refraction had not yet been formulated.

For something more specific on the character of Bacon's background knowledge, we offer the following summary of his "Optics":

1. He differentiates physical vision from "higher spiritual vision."

2. He divides his physical vision into "*direct* vision, *reflected* vision, and *refracted* vision."

3. Direct vision is handled by analysis and diagrams of the composition of the eye, including the lens, the optic nerves, "their relation to the brain and the coats and humors of the eye." For vision, however, Bacon emphasizes that there must be, in addition to an eye, "impressions or species emanating" from the visible object. The object must be in front of the eyes, as contrasted, for example, to hearing or to smell, and must be of suitable size and density. A medium must exist for the "impressions or species" to travel to the eye, and must be of "rare density." "Mixture of color rays" occurs both in the pupil of the eye and in the medium. Difficulties, errors, or optical illusions in vision are due to such factors as smallness of pupil, state of health of the eye, oblique angles of vision as "in producing double images," "rays proceeding from both eye and object" (Aristotelian influence here), excessive distances, and rapidity of motion of object or of observer (advance shades of Doppler's principle here). (One could devote an entire paper to penetrating Bacon's handling of the material above, for example, his extensive analysis of what Aristotle and Seneca originally labeled as "weaknesses of the human eye." Neither time nor the objective of this chapter will permit more than this summary of Bacon's optical knowledge as it reflects his times.

4. Bacon's second category of vision, "reflected vision," revolves primarily around empirical geometrics of mirrors already well known to the Greeks and Romans, and with diagrammatic and general emphasis upon the general effects of the shape of the mirror—that is, plane, concave, or convex, spherical, conical, or cylindrical—upon the shape and "the place of the image."

5. In his treatment of "refracted vision" Bacon shows such knowledge as:

   a. Reflection requires "a two-field medium."

   b. Refraction requires an "oblique angle," that is, one in which the ray is "not perpendicular to the surface of the second medium."

   c. Refraction takes place at the plane, concave, or convex surfaces, including the eye lens.

d. The general direction, diagrammatically and empirically, of the bending of refracted rays occurs according to the character of the lenses or relative densities of the two mediums.

e. The broken appearance of a stick in water and various magnification effects of moisture, lenses, etc., may be explained as refractive phenomena.

BACON'S GENERAL METHOD AND THESIS IN THE RAINBOW

Bacon's detailed treatment of the rainbow occurs not in his general section on "Optics" in his *Opus Majus,* but in the succeeding section on "Experimental Science." This is not accidental: Bacon specifically introduces the rainbow in order to illustrate what he considers the scientific method or approach. He emphasizes the method of science to be that of "the test of experience." In dealing with natural as contrasted to supernatural phenomena, Bacon essentially calls for an inductive approach: he suggests a way of proceeding, without preconceived ideas, directly to phenomena and, in all possible instances, of examining it with apparatus or devices especially useful for the purpose. Consequently, in using an investigation of the rainbow as an example of this essentially inductive process, Bacon, in general, is not defending, justifying, or testing a thesis or hypothesis formed in advance. He makes a point of attempting specific projects out of his study of such aspects as altitude, magnitude, shape, color, time and place of occurrence, the reflective or refractive nature of the rays involved, and the forming of conclusions based on these projects.

Consequently, it becomes of particular interest to note that in spite of his knowledge of refraction and his inductive method, Bacon's conclusions are often very close to or parallel with those of Aristotle and Seneca. In fact, Bacon eventually concludes that refraction is not a factor in rainbow formation; he agrees with Aristotle and Seneca that the rainbow is a reflected image of the sun. However, in making a principle out of his direct work with data, Bacon does come up with a great deal more firsthand detail than Seneca. And he is, of course, doing his investigation twelve centuries later.

BACON'S METHODS IN MORE DETAIL

As a concrete example, then, of his method of "experimental science," Bacon first attempts to study under what conditions rainbow phenomena occur. He does this, as he emphasizes, not merely "by discussing the phenomena" but by seeking out actual visible objects and situations having the same colors and form as the rainbow.

1. His "experimenter holds a number of transparent and dark hexagonal crystal stones in a solar ray through the window to bind all the colors of the rainbow in the shadow near the ray."

2. The experimenter or observer in seeking other situations finds them in water falling from rowers' oars or from mill wheels or during a rain—but in all cases, he notes that the sun is shining. Also, the experimenter, like Seneca's fuller, "vigorously sprinkles water from his mouth into the sun's rays while he is standing alongside of the rays." The phenomenon is also seen around a candle at night or through moist, half open eyelids or upon oil surfaces. In every case there is an identical factor noted: the sun or another light source.

3. Bacon is not yet, however, attempting explanation. He is still seeking data. The crystalline stones, the moist eyelids and eyebrows, dewfall, and lighted candles as well as slitted rays are used again, this time to examine the shape of the colors. Bacon notices, for future reference, that the crystalline stones show rainbows of "straight shape" while the other cases reveal complete or nearly complete circles.

4. Next he turns for data to the rainbow of original interest, the one in the heavens. This time he is interested in taking a series of angular or altitude measurements above the horizon of both the sun and the rainbow, in order to determine their position relative to each other. By means of an astrolabe he finds that:

   a. The higher the altitude of the sun, the lower the bow and vice versa.

   b. The rainbow is always opposite the sun with a line passing from the center of the circle of the bow back to the center of the eye of the observer, who is facing the bow.

   c. The maximum elevation of the rainbow is found to be 42°, a value found when the sun is at its lowest point, on the horizon at sunrise or sunset.

   d. The value of 42° is also found by measuring the altitude of the sun when a trace of the blue of the rainbow can just be seen at the horizon. Results *c* and *d* represent the two limiting conditions that become a check upon one another.

With the above data Bacon accounts for Aristotle's and Seneca's observations that the rainbow does not appear in the heat of a summer day. Because of the limiting angles of an elevation of 42°, rainbows cannot appear when the sun is above 42°. Thus, we find for the first time that Bacon is offering an explanation, operationally determined in terms of inductive experimental data.

The moral here, of course, is *measurement*. Since the astrolabe was already known in antiquity, it is not in the device itself, but in the *emphasized principle of all possible uses of such devices in direct problem solving* that we, in this study of method in the study of the rainbow, find Bacon's real contribution.

5. Bacon's next project is investigation of the size and shape of the rainbow. For the purpose he uses a geometrical model, a cone, the apex of which would rest on the semicircle of the bow, and the axis of which would be the above-mentioned line from the eye to the center of the bow. Since the center of the bow by the above measurements with the astrolabe is also always in a line with the center of the sun and the sun's nadir, the base of the cone would be elevated and lowered with the lowering or elevation of the sun. Having established this cone as a mathematical model to fit the phenomena of the actual rainbow in the skies, Bacon then applies it to explain the size and shape of all circular spectra as well as the contradiction that baffled Seneca regarding full circles versus semicircles. He accomplishes this by showing geometrically that if the base of the cone is elevated and depressed inversely with the sun:

a. Short cones formed with the eye by nearly falling drops of water as a base give an entire circle to that base.

b. Long cones formed by more distant rainbows, as in the customary phenomena of the heavens, means extension of the cone until it touches the earth, with a consequent limit for appearance of the entire circle as a base.

c. The more of the base that is cut, the less that appears as rainbow.

d. Therefore, in the limiting case of a rising or setting sun, the 42° maximum elevation of the rainbow is geometrically not enough to provide a full circle or base to the cone.

e. Higher positions of the sun means the cutting of more and more of the base of the cone until beyond the 42° elevation of the sun, none of the base of the cone representing rainbow phenomena can be seen.

Thus, by the employment of a mathematical model to fit phenomena, Bacon is taking advantage of a knowledge tool used successfully in modern statistics as well as modern and classical physics. Perhaps in Bacon's case, the success of this geometrically descriptive model in explaining contradictions in the shape of the rainbow phenomenon may have been a factor making it unnecessary for him to use refraction in explaining the rainbow. Filling in gaps as above and resolving dilemmas not quite overcome by Aristotle and Seneca, enabled the continuation of a mirror reflection theory as the simplest possibility. However, as we shall see, Bacon does later pose the question of reflection versus refraction and attempts to put it to a test in connection with the rainbow.

6. From explanation of the size, shape, and diversity of the bow, Bacon applies himself to the questions, in more detail, of time and place of rainbow formation. This he does by analysis of the established facts of daily, seasonal, and latitudinal variation in the altitude of the sun. That is, by geometrical considerations, in terms of degree, of daily, latitudinal, and seasonal variation in the position of the sun, corresponding positions or possibilities for the bow can be determined. This, of course, involves considerations of the greatest daily altitude of the sun at noon; the greatest seasonal altitude, as the summer solstice in Northern regions above about 23.5° latitude; or the greatest latitudinal altitude, as the sun's zenith at the equator during the spring and fall equinoxes.

Narrowed down, this means that the position of the bow can be predicted at any given point or latitude through calculations or considerations based on facts of:

    a. the relationship of the bow to the sun's motion in a daily circle,

    b. the relationship of the bow to the sun's circle of altitude due to change of season, and

    c. the nonappearance of the bow with any altitude of the sun above 42°.

I believe the above speaks for itself as method involving use of facts in established mathematical relationships for benefits of prediction.

7. Next Bacon seeks to determine "whether bows are seen by incident, reflected or refracted rays." This question is a crucial one, determining whether new knowledge of refraction—new in respect to Seneca and Aristotle—plays a part in understanding the rainbow or whether former mirror concepts remain sufficient. To Bacon, the crucial test in resolving this problem is to determine whether or not a rainbow remains fixed, or varies in number or position according either to the number or motion of observers.

    a. He identifies a stationary bow, that is, one remaining fixed irrespective of the motion of an observer, with either an incident or a refracted ray. The very nature of incident rays going out in different directions from a rainbow source and witnessed by an observer, would permit the rainbow to be seen in only one place no matter what the position of the observer. On the other hand, Bacon's attempted evidence that refracted rays would also result in a fixed image are based on claims of refracted images seen in the same position by observers in different positions. He uses the crystalline stones with their rectangular rather than circular rainbows as well as a fixed stick protruding from water as cases of refraction showing fixed images irrespective of the position of the observer.

b. He identifies reflected rays with a "separate bow for separate observers or with a bow that would move in respect to a moving observer," on the basis of what happens to images in actual cases of reflection from a mirror.

c. In Bacon's crucial test for alternatives *a* or *b* above, two observers moving in opposite directions involve effects of rainbows moving in opposite directions. Since this effect of "a bow for each observer" as well as of "a bow moving parallel to the observer" have previously been identified with *reflected* as opposed to incident or refracted rays, Bacon concludes that rainbows in the heavens as against those of crystalline stones are due to reflected rays.

How does it happen that Bacon came to the wrong conclusion on this crucial question of refraction? Or, to put it another way, how was it that with the benefits of his knowledge of refraction, and of his emphasized principle of inductive, empirical testing, Bacon eventually emerged with a reflection theory of the rainbow very similar to that of Aristotle and Seneca? As I see it, the answer lies chiefly in two incorrect assumptions, one in his interpretation of relative motion of images in refraction, the other in what I would refer to as an incorrect *either-or* assumption in *a* and *b* above. In the first case, because in the motion of an observer, the part of the stick in the air appears to change position relative to the part in the water, he assures that the submerged part, which, of course, is a refracted image, is fixed, that is, it would appear in the same position to different observers. This in turn gives Bacon a false basis in *a* and *b* above for differentiating between a reflected and a refracted image. Thus, while reflected and refracted images can be differentiated in terms of different relative motions to a moving observer, a basis of differentiation would not be that of a "fixed" refracted image as interpreted through a false assumption by Bacon. Second, and perhaps even more important in the history of science or thought, there was this *either-or* assumption. Even if he had been correct In differentiating between refraction and reflection on the basis of a "fixed" refracted image, by setting up reflection *against* refraction, Bacon would not have been allowing for a possible third alternative of combined effects in the rainbow of reflection *and* refraction. Both *do* exist in the complete picture of the rainbow. In my opinion this *either-or* assumption must even today be continually guarded against. Perhaps the full roles of heredity and environment, for example, can best be understood not when the problem is posed as heredity versus environment, but as the interplay of heredity *and* environment. Our greatest benefits in modern optics have been and are to be obtained not by canceling out conceptions of light as electromagnetic waves in favor of light as quantum particles, or vice versa, but by considering both as different aspects of a same thing.

A third source of error on Bacon's part, which tied up with the other two, was the incompleteness of the "broken stick" as a refraction analogy to what actually occurs in the rainbow as a combined phenomenon of reflection and refraction. We mention this in order again to emphasize the dangers of analogy and the constant care to be taken in its use.

8. Having thus satisfied himself with what he considers a crucial test of the reflective rather than refractive character of rays from the rainbow, Bacon expresses agreement with Seneca's general mirror hypothesis. That is, each raindrop becomes a mirror. Since the drops are in close proximity, the combined images assume the appearance of a continuity. As previously explained, however, Bacon completes Seneca's general mirror conception with his own concept.

9. Bacon's concluding sections are concerned with the color of the bow.

    a. He points to his former evidence that a "bow changes with a change in observers," i.e., with a change in the position of an observer, as a basis for his conclusion that the colors of a bow do not actually exist in the rainbow, but in the sensation of the observer.

    b. Unable to explain color diversity empirically, Bacon does so metaphysically. He refutes Aristotle's and Seneca's explanation that variation in color is due to variation in cloud density by referring (1) to the variable colors seen when examining the crystalline stones which do not have variation in density, and (2) to the rainbow in sprinkled drops of water, likewise uniform in density. So far so good. Then, speculatively analyzing colors of the rainbow into five primary hues of red, blue, green, white, and black, Bacon attributes the presence of these colors in the rainbow to "the number five being in nature better than all other numbers, as Aristotle says in the book of Secrets," and therefore to the explanation that "because the number five distinguishes things more definitely and better, nature for this reason rather intends that there shall be five colors in the Rainbow." (If I may digress a moment, Bacon in his teleology here is perhaps typical of mankind in general who, while pushing back the boundaries of knowledge by means of the empirical, will, at any given time, at particular temporary limits of the empirical, recess into the receding, mystical beyond.) At any rate, Bacon's concluding sections on color are clearly his weakest, particularly the criterion of the inductive pattern of projects he previously set up in both principle and practice in previous sections. Most likely he was not aware of any basis for specific projects for color. That course became Newton's great contribution in optics, but as we shall see, only after first being on the track of refraction. Thus, Bacon's problem solving with the rainbow reached its peak short of the point where he incorrectly rejected refraction as part of an explanation of rainbow formation. The best of methods can be handicapped by incorrect or insufficient material with which to work.

Bacon's investigation of the rainbow has been examined in enough detail here hopefully to reveal the high points of his problem solving as well as the general empirically inductive character of his method outlined in our overview.

THE OPTICAL KNOWLEDGE BACKGROUND OF NEWTON'S RAINBOW

Isaac Newton's explanation of the rainbow as revealed in his own treatise on "Opticks"[4] has, besides knowledge of the laws of reflection, a threefold basis:

1. Snell's laws of refraction establishing earlier in the seventeenth century that

    a. refracted and incident rays are coplanar with the normal drawn through the common point at the surface, and that

    b. $\dfrac{\sin i}{\sin r} = n$ where sin i = sine of angle of incidence

    sin r = sine of angle of refraction
    sin n = a constant or index of refraction.

2. Archbishop Marco Antonio de Dominis's and René Descartes's independent applications of Snell's laws as well as general laws of reflection to rainbow phenomena to explain both bows. Since Newton incorporated the work of Dominis and Descartes into his own, I shall explain this in more detail later.

3. Newton's own work with prisms in connection with color dispersion of sunlight. Since this was indispensable to understanding rainbow color, we shall also examine this in detail further on. I shall rarely point out here that in his work on dispersion, Newton was the first thoroughly to realize the dependence of Snell's constant or index of refraction upon the nature or color of light employed. This was what was needed to unlock the mystery of the colors of the rainbow, as we shall see.

NEWTON'S GENERAL METHOD

Newton's general method is immediately obvious and definite. This is nowhere better shown than by Newton himself in the table of contents of book 1 of his *Opticks:*

BOOK ONE

PART I.	Definitions	I–VIII
	Axioms	I–VIII
	Propositions	I–VIII
PART II.	Propositions	I–XI,

and his terse one-sentence introduction: "My design in this book is not to explain the properties of light by hypothesis, but to *propose* and to *prove* them by reason and experiments: in order to which I shall premise the following definitions and axioms" (italics added).

In other words, Newton is establishing in this book, as in other of his scientific works, a particular pattern of scientific method which became associated with his name in both classical and modern physics as well as in other areas of science. It is a hypothetico-deductive pattern or system in which, in scientific problem solving, the best possible hypothesis in terms of insight or of past experience is proposed for acceptance or rejection by controlled experimentation. As a method, it is the dialectic opposite to the inductive principle used by Roger Bacon with the rainbow. Rather than waiting for a hypothesis until an investigation of facts has been made, a "best" hypothesis is formed first in order to be specifically tested as either tenable or untenable. However, in setting up such propositions, as is obvious from his table of contents as well as from further examination of the book itself, Newton is careful and precise: First, through definitions he defines his main terms; second, through axioms, he makes clear the basic premises and assumptions upon which his propositions are based.

Thus, the following turns out to be the general content and method of Newton's optical investigations:

1. He defines his uses of such terms or expressions as rays, refrangibility, reflexibility, angle of incidence, angle of reflection, angle of refraction, sine of incidence, sine of reflection, sine of refraction, homo- and heterogeneal light.

2. His eight axioms, with descriptions and diagrams, turn out to be the laws of reflection and Snell's laws of refraction described above together with the mechanics of their application in mirror, lens, and prism-image formation. In these axioms we find use of terms defined above.

3. His propositions I–VIII represent, outside of his corpuscular theory, his outstanding work in optics. They involve:

    a. hypotheses of various aspects and under various conditions in which Snell's n, the index of refraction, varies with the nature or color of the light used, and

    b. the famous sixteen diversified prism experiments as evidence in the acceptance of the above hypotheses.

4. Propositions I–XI of part 2 of Newton's book turns out to be further concepts, applications, and experimental verifications leading from the basic eight propositions of part 1.

NEWTON'S TREATMENT OF THE RAINBOW ITSELF

Newton's treatment of the rainbow is proposition IX or problem IV of the second group of propositions described above titled, "By the discovered properties of light to explain the colors of the rainbow." Of direct significance to this proposition of the rainbow are two previous propositions of the first group: Prop. II. Theorem II: "The light of the sun consists of rays differently refrangible," and Prop. III. Theorem III: "The sun's light consists of rays differing in reflexibility, and those rays are more reflexible than others which are more refrangible." I shall, therefore, first turn attention to these previous propositions and to the eight prism experiments for their verification.

Treatment of these propositions in obviously Newtonian style involves first a statement of the proposition, followed by detailed accounts and diagrams of the experiments and the reasoning involved under the heading of "The Proof by Experiments." I believe that I can best reveal Newton's methods in these by systematically outlining them in their continuity, especially since Prop. III, given above, is verified by the last two experiments of Prop. II.

PROPOSITION II. PROOF BY EXPERIMENTS

First Experiment. To Obtain Systematic Measurements of the Sun's Image through Various Apparatus

Method of Experiment

1. "In a very dark chamber, at a round hole, about one third of an inch broad, made in the shutter of a window, I placed a glass prism, whereby the beam of the sun's light, which came in at that hole, might be refracted

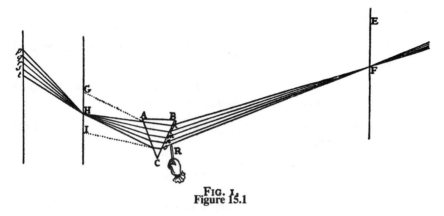

FIG. 1.
Figure 15.1

Source: Isaac Newton, *Opticks* (New York: Dover, 1952; based on the 4th ed., London, 1730).

upwards toward the opposite wall of the chamber, and there form a colored image of the sun."

2. For dispersive power control in this and all prism experiments, Newton then adjusted his prisms for a prism axis perpendicular to the incident rays as well as for a minimum angle of deviation of the entering and departing rays.

3. Recording was then made of the

   a. sun's apparent diameter by measurement at the prism of the subtended angle* of the incident beam,

   b. refracted angle of the prism,

   c. length and width of the image,

   d. distance of the image,

   e. subtended angle of the beam to the image.

4. Steps 1–3 were repeated five times as a check.

5. Steps 1–3 were repeated with other prisms of other refracting angles as well as of other refracting materials such as water between glass panes.

6. Steps 1–3 were repeated with different sizes of hole, different inclined angles, and different passageways through the prism.

---

*The subtended angle is the angle of the refracted ray.

236   Part Three. The Human Dimension in Science

Results

1. The image of the sun was not circular but about five times longer than wide in each case.

2. The subtended angle of rays leaving the prism were in each case about five times larger than the subtended angle entering.

3. The image under the varying conditions above always retained the same measurements.

Reasoning and Conclusions

1. Since the prisms were set for the refractions on both sides of the prism to be equal to each other, by ordinary laws of refraction, Snell's laws, there should be the same angles subtended on both sides.

2. With the noted results of the subtended angles on the side leaving the prism being about five times as large as on the entering side, there is involved a breaking up of the original ray entering at one angle of incidence into various angles of refraction and causing an image that would ordinarily be round to become almost rectangular.

3. At positions on the image associated with varying angles of refraction are varying spectrum colors.

4. Therefore, the experimental proposition II above is supported.

Second Experiment

This experiment is really an extension of the first experiment, but with Newton this time looking through the prism at the round hole in the window.

Method and Results

1. The prism axis is again turned perpendicular to the beam.

2. The prism is also again adjusted for a minimum deviation angle: that would be the position where upon "looking through the prism upon the hole, and turning the prism to and fro about its axis, to make the image of the hole ascend and descend," a stationary point would be reached "between the contrary motions" of the image.

3. Newton then examined, through the prism, the round hole in the window.

4. Result:

   a. The image in the round hole seen in this way was again many times longer than wide.

   b. The spectrum seen was such that red at one end was the color least refracted while violet at the other end was refracted the most.

Conclusions

Refraction of the colors does not take place according to one proportion of the angle of incidence to the angle of refraction. A considerable inequality of angles of refractions from the same incidence angle occurs.

Thus, the first two experiments together identified particular colors with particular angles of refraction and particular indices of refraction.

Third Experiment to Determine the Effects of Using Two or More Prisms Together

Method

This is essentially an extension of the first experiment, but with the use of two, then three, and then four prisms in combination.

Logic of the Use of More than One Prism

If the image with one prism was made oblong by "dilatation" of the ray rather than by the separate refraction effects claimed by Newton, a second prism should increase or decrease the "dilatation," and the width of the image should be correspondingly increased or decreased. obviously, by this third experiment, Newton was anticipating other possible interpretations besides his own conclusions of separate refraction effects for each color in the former two experiments. This "dilatation" of the original incident rays was a possible alternative.

Results

1. The breadth of the image was not increased, but merely rotated by the various combinations of prisms.

2. The order of colors always remained the same.

Conclusions

Newton's former conclusions as to separate refraction effects for each color are further confirmed.

Fourth Experiment, to See Whether If Separate Sections of the Refracted Ray were Independently Sent through a Second Prism, They would Maintain Their Relative Refractibility

Method

As already shown in Figures 12.4 and 5, each color in the spectrum from the first prism was separately permitted through a second prism. In each case, the further refraction was recorded.

Result

Each color or section maintained its relative refractibility with red bending the least and violet the most.

Conclusion

Separate refraction effects for each color are further confirmed.

Fifth Experiment. To Further Test Separate Refraction Effects of Different Colors by Separation of Colors Previously Blended

Method

1. Newton blended two isolated colors, red and violet, to get purple.

2. Through another prism, Newton looked at the purple mixture and found the violet and red separated, with the violet, as usual, further refracted than red.

3. To show that the final images were from the original two colors, he interfered with the passage of the original violet, and the final violet disappeared. Likewise with the red.

Conclusion

This provided further confirmation of the previous conclusions.

Sixth Experiment. To Further Test Separate Refractive Effects according to Color through Variation of Focal Points through a Lens

Method

Newton used a lens with different sections of colored light from a prism in order to determine the distance of the image by various colors.

Logic of the Method

If red is refracted less than violet, it should come to a focus at a point farther from the lens than the violet.

Results

The red *did* converge at a point farther from the lens than the violet.

Conclusion

This provided further confirmation of the previous conclusions.

Ninth and Tenth Experiments. To Test, in Addition to the Previously Tested Proposition II, Proposition III "That the Sun's Light Consists of Rays Differing in Reflexibility, and that Those Rays are More Flexible than Others Which are More Refrangible."

Ninth Experiment

1. In the usual prism setup, Newton turned his prism through the critical angle, and found that the rays of greatest refraction were totally reflected first.

2. By then adding a second prism and a paper screen to catch reflections from the first prism as they came through, he found that at first there was more violet and blue reflected than green and red, until the latter became totally reflected in the first prism, and more intense on the screen placed after the second prism.

Tenth Experiment: Extension of the Ninth

1. In this experiment, whether Newton refracted a given light through the two prisms or afterward arranged to have the light totally reflected by one prism and refracted by the other, there was no difference between the "degrees of refrangibility" of the same colored light.

2. Here again the colors with the greater degree of refraction totally reflected first.

Conclusions

1. Further confirmation of Proposition II.

2. Verification of Proposition III.

NEWTON'S THESIS OR PROPOSITION ON THE RAINBOW

With the detail that I have already given on Newton's general scientific approach as well as on the specific relevant experimentation, it should now be evident that by the time he reaches the proposition of the rainbow, he has so well and so systematically cleared the ground that the rainbow becomes a particular application of his previous confirmations of dispersion. In this case, of course, it becomes a dispersion through the refracting and reflecting medium of raindrops, with full detail as to relative positions of sun, raindrops, and observer.

NEWTON'S SPECIFIC TREATMENT OF THE RAINBOW PROPOSITION

His treatment has a twofold emphasis:

a. that involving *the origin of the primary and secondary bows themselves* rather than their colors, which he attributes to independent geometric explanations and the experimentation of Dominis and Descartes. Their thesis involves two refractions and one reflection in the primary bow and two refractions and two reflections in a secondary bow, and

b. that involving his own contributions of the *origin of the colors of the bows* in term of dispersion of the component colors of "white light."

Newton had thoroughly investigated the work of Dominis and Descartes, and had found their work valid with the efforts of one independently check-

ing those of the other. Newton had also already established his own propositions of dispersion, as we have seen. Consequently, his chapter or proposition
necessitates merely a synthesis of his own and previous contributions of
Dominis-Descartes. The synthesis Newton accomplishes to explain the rainbow is still valid today. The following is a summary of his synthesis.

For a rainbow, three conditions are necessary as Seneca himself knew: (1)
the presence of water drops in a cloud or mist, (2) the sun shining on such
drops, and (3) an observer with his back toward the sun and his eyes toward
the mist or clouds.

As illustrated in Figure 15.2, however, rays from the sun may form two
bows, the brighter primary and the fainter secondary, which is not always seen.
If both are visible, the secondary is above the primary and has its colors in
reverse order. By tracing through, on the figure, light rays from the sun, it can
be seen that given rays forming a primary bow experience *two* refractions or
bendings—one on entering and one on leaving the drops—as well as *two*
reflections within the drops. On the other hand, in the case of the primary bow,
there is only *one* internal reflection with the two refractions. Since some light
is lost with each reflection, it becomes clear why the primary bow is brighter
than the secondary, or why the secondary cannot always be seen.

In being refracted, the ordinary white light from the sun is separated into
its constituent parts of red, orange, yellow, green, blue, indigo, and violet
because of the different refractive angles of these different colors. In order to
see the bows, the eye E must have a position relative to the sun and to the drops
such that when the rays are refracted and dispersed by the drops, a definite
angle is formed with the line AE parallel to the sun's rays. This angle varies
from color to color since the index of refraction varies according to the refractive angles of each color. In the primary bow, the angles vary from 42° in angle
REA for the red, to 40° in angle *VEA* for the violet. In the secondary bow, the
angles vary from 54° for the violet to 51° for the red. The explanation for the
semicircular shape of the rainbow involves the focus of a point representing
a given color, say red, In the primary bow, that would form the 42° angle *REA*.
The locus of such a point would be half or part of a circle forming part of the
base of a cone, with the eye at the vertex of the cone as previously noticed by
Roger Bacon but not explained until Newton. As shown in the diagram by a
supplementary small arrow from each drop, rays of various colors are leaving
each drop, but for each drop in a given section of the bow, because of the differences in refractive and reflective angles, only the ray of the particular color
seen strikes the eye. The other rays from the same drop or drops fall long or
short. This, along with the fact that the secondary bow, in addition to two
refractions, has two reflections while the primary bow has one reflection, is the
explanation for the secondary bow having a reversed order of colors. This is
made clear in Fig. 15.2.

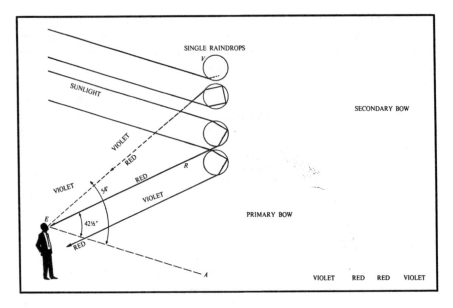

Figure 15.2

Rainbow formation. Drops of water refract, reflect, and disperse sunlight. To see a bow, an observer must face a mist and have his back to the sun.

GENERAL SUMMARY

We have attempted in this chapter to examine the scientific problem-solving techniques of Seneca, Roger Bacon, and Isaac Newton in the case of the rainbow. In order to do so intelligently, in each instance, we have considered the optical knowledge background of each investigator and consulted the original sources. We have emphasized the investigator's general methods, summarized in our preliminary overview, as well as his specific techniques and methods that in most instances revealed the larger pattern of his approach. In each case, in addition to method, as we went along we pointed out specific contributions of the investigator to conceptions of the rainbow itself, as Seneca's three factors of origin, Bacon's cones, and Newton's dispersion of component colors of "white light."

We also found that the scientific treatment of the rainbow advanced, like science in general, from haphazard observation by Seneca to organized inductive investigation by Roger Bacon to controlled hypothetico-deductive experimentalism by Newton.

We have seen how creative partial pictures of the rainbow changed in time with accumulated knowledge, experience, and techniques. Here again creative scientists appear as interactors in and with nature rather than as merely detached observers. The rainbow is a product of this interaction. It starts out as sensation, becomes conceptualized, and forms changing partial pictures of itself with advances in experimental techniques.

NOTES

1. Lucius Annaeus Seneca, *Naturales Quaestiones* (New York: Macmillan, 1910).
2 Aristotle, *Meteorologica* (Oxford, England: Oxford University Press, 1923).
3. Roger Bacon, *Opus Majus,* vol. 2 (Philadelphia: University of Pennsylvania Press, 1928).
4. Isaac Newton, *Optics* (New York: Dover; based on the 4th ed., London, 1730).

SUGGESTIONS FOR FURTHER READING

Cohen, J. Bernard. *Introduction to Newton's* Principia. Cambridge, Mass.: Harvard University Press, 1971.
Encyclopaedia Britannica, 11th ed. Chicago: University of Chicago Press, 1910. For a general history of science background materials to 1910, this edition of the *Encyclopedia* is perhaps the most scholarly.
More, Louis Trenchard. *Isaac Newton.* New York: Dover, 1962.

16

Measurement as Interaction: Space, Time, and Einstein

Space by itself and time by itself are doomed to fade away into mere shadows and only a kind of union of the two retains self-independence.

Herman Minkowski, public lecture, 1908

ABSOLUTE SPACE AND TIME

Newtonian space was a space of common sense. Like a huge container, it housed all things; absolute, it existed independently of events. Newton wrote in his *Principia,* "Absolute space, in its own nature, without regard to anything external, remains always similar and immovable." Time was absolute, too. "Absolute, true, and mathematical time, of itself and from its own nature, flows equably and without relation to anything external, and by another name is called duration." Time rolls relentlessly on. Robert Burns had the idea when he said, "Nae man can tether time or tide." Or, as Chaucer expressed it in his *Canterbury Tales,*

> For though we sleep or wake, or roam or ride,
> Ay flees the time, it will no man abide.

If space and time were the settings for all things, they had a commonness, a likeness for everybody. Here, there, and everywhere a second was always a second, a mile always a mile. Space and time were also continuous: space was a three-dimensional succession of points, and time a separate one-dimensional succession of instants. Actually, no gaps existed in the continuity of points or instants. And there were no first or last points in space or time. The series of points was mathematically infinite. Space was reversible in direction: it extended right or left, up or down, backward or forward. Time, however, was not reversible. It flowed only one way—forward.

245

Newton, like the ancient Greeks and Romans, recognized a universe of motion. Space and time were absolute frameworks in which the motion of all things could be measured. Standard metersticks and clocks guaranteed the same measurements for everybody. Newtonian mechanics, based upon absolute space and time, did very well for over two centuries; indeed, modern technology sprang from it. For all immediate practical purposes, the earth could be treated as stationary, with all objects at rest or in motion upon it. Of course, Huygens's wave theory of light required an unseen, weightless ether as a medium in space for propagation of light waves. But when a fixed ether was necessary also for Faraday's fields and Maxwell's electromagnetic waves, the ether too seemed firmly set in Newton's mechanical universe.

At the end of the nineteenth century, scientific progress seemed but a matter of improved techniques, of greater precision of measurement. Scope, structure, and standards of operation were thought to have been fully achieved. Little did scientists at the time realize that their time-space experiments were leading to an explosion that would rock the physical sciences to their foundations.

THE GREAT DILEMMA

In 1881, A. A. Michelson and E. W. Morley tried to "prove" that an ether exists by means of an ingenious device called an interferometer.

A man can row across a river and back sooner than he can row the same distance D upstream and downstream. Suppose the current is 1 mph and the oarsman can race at 5 mph in still water. Upstream, he therefore travels at 4 mph, and downstream, at 6 mph. At first glance, the loss of velocity against the stream seems to be balanced by the gain downstream for an average of (4 + 6)/2, or 5 mph. The.average is incorrect. Actually, the boat spends *more time* traveling half the distance *at the lower speed* (4 mph) than traveling the other half at 6 mph. The rower's average speed, therefore, will be somewhat less than 5 mph (4.9 mph). The boat moves across the current and back faster than upstream and downstream.

In the interferometer, Michelson and Morley substituted a light beam for a boat and an ether "stream" for a water current. If the earth moves in an ether, there will be a relative ether "stream" opposite and equal to the earth's motion. By analogy, a relative wind is felt on a hand extended from a moving car even in otherwise still air. From the earth's revolution alone, there should be an ether current of 19 miles per second. A beam of light is split in the interferometer, and each section sent at right angles through a system of mirrors. The split beams travel equal distances and are reflected back to the same telescope. A beam in the direction of the earth's motion would first move against the ether "current" and then with it. The other beam would be across the ether "current."

The analogy is therefore complete: the beam across the ether should complete the round trip before the other. As the two beams are no longer in phase, a changing pattern of light should be observed at the eyepiece. The apparatus was sensitive enough to detect a difference of only 2 miles per second in the speed of the two beams.

Michelson and Morley repeated their experiment many times at different angles of rotation and at various times of the day and year. However, the predicted result did *not* occur; the light intensity in the telescope did *not* change as expected. This meant that the velocity of the light beams remained the same regardless of directions. No ether "stream" effects existed. Whether the earth moved toward or away from a star, light from the star would still have a speed of 186,000 miles a second with respect to the earth. Suppose car *A* at 80 mph speeds past car *B* moving at 50 mph on a freeway. At constant velocities the two cars would be either 30 or 130 miles apart in an hour, depending upon whether the cars move in the same or opposite directions. In the case of a light ray passing the earth, whether the earth moves with the ray or against it, the ray is 186,000 miles away at the end of a second (Fig. 16.1). Velocities just did not seem to add up. The Michelson-Morley results violated the principle of addition of velocities at the heart of mechanics, thus presenting a seemingly impossible dilemma. What was to be done?

UNIFYING THE ETHER KNOT

The ether at the center of the Michelson-Morley experiment had not been detected. Possible solutions to the problem were that (1) both the earth and the ether were stationary, (2) the ether moves with the earth, (3) the earth partially "drags" the ether, (4) the apparatus shrinks in the direction of motion, or (5) ether does not exist.

Established ideas are not easily discarded. The ether was a basic assumption of the mechanical view of the universe; to give up the ether was to weaken the foundation of the universe itself. A stationary earth could save the ether, but nobody cared to return to the Ptolemaic system. To say that the entire ether moved with the earth was also earth-centered and therefore indefensible: the outermost reaches of the ether would have to move preposterously fast to rotate daily with the earth.

An ether drag concept like that of French physicist August-Jean Fresnel (1782–1827) seemed more feasible. Perhaps the moving earth dragged some ether in its wake much like a ship drags water. In ether drag, light from a given star would be carried with the ether and could come in from the same direction at all points in the earth's orbit. But unfortunately, in 1728, the English astronomer James Bradley had observed that the direction of starlight *does* vary

Figure 16.1

The absolute speed of light. Whether the earth moves toward or away from a star, light from the star has a speed *c* of 186,000 miles a second with respect to the earth.

with the earth's position in its orbit much as raindrops on the windshield of a moving vehicle. This *aberration of light* eliminated the ether drag hypothesis.

A more significant alternative to save the ether was the Fitzgerald-Lorentz contraction hypothesis (1893), according to which all objects are shortened in the direction of their motion in the ether. A ball of clay hurled to the ground flattens; why not moving objects opposed by ether? A light ray moving in an interferometer arm against the ether may very well be slowed down by the ether stream. But that light ray would have less far to go if the interferometer arm shrank. The contraction of the arm distance could balance out the lower velocity of the light ray, and both rays could return to the telescope together. The Irish physicist George F. Fitzgerald (1851–1901), who introduced the contraction hypothesis in 1893, even developed the following equation to find the necessary contraction of the interferometer arm:

$$l = l_0 \sqrt{1 - \frac{v^2}{c^2}}$$

where l_0 = the original length of the arm
   l = the length when contracted
   v = the apparatus velocity in the ether
and c = the speed of light.

Note that if $v = 0$, the term $\sqrt{1 - v^2/c^2} = 1$, and $l = l_0$. No apparatus motion, no shrinkage. If $v > 0$, the factor $\sqrt{1 - v^2/c^2}$ is less than 1, and l is a fraction of l_0. Shrinkage of arm length to l depends upon apparatus velocity v in the ether.

Ingenious indeed! Epicycles had reconciled Ptolemaic theory with the observed positions of the planets; now contraction of moving substances was being suggested to reconcile classical mechanics with the Michelson-Morley results. But Fitzgerald could not specifically explain what causes objects to contract in an ether or why all substances—steel, wood, glass, or copper, for example—would flatten equally. In 1895 the great Dutch physicist Hendrick Lorentz (1853–1928) came to Fitzgerald's rescue with an electrical theory of matter: Matter contains electric charges that exert strong forces on each other, he said. These charges are accompanied by electromagnetic fields. An object moving in the ether contracts because the charges and fields within the object interact with the ether stream. A new electrical view of matter was assisting the mechanical view, but Lorentz's specific field-ether interaction could not be verified. The Lorentz-Fitzgerald contraction hypothesis rested on uncertain grounds, but it achieved new meaning with Einstein's relativity.

Einstein with one stroke cut the ether knot: he *discarded* the ether. All velocities are relative. Nobody knows the absolute velocity of anything. The fixed framework of an assumed ether cannot be detected; the ether, therefore, is meaningless and unnecessary. Einstein was aware that the mechanical view had run into a number of serious contradictions. Electromagnetic forces between objects worked at right angles rather than in a line of attraction or repulsion. Mercury's orbit showed a strange rotation of about 43" of arc per century not explainable by Newton's gravitation. And in 1901, W. Kaufman and J. J. Thomson found that electrons increase in mass the faster they move. In Newtonian mechanics, mass was a fundamental constant of the universe. How could mass increase? What was wrong? To Einstein, classical mechanics had reached its limits. The ether dilemma was only one item in the decline of the mechanical view. The ether was an invention of the mechanical view to transmit electromagnetic waves that Einstein's photon theory did not need. In abandoning the ether, Einstein rejected the mechanical view behind it and built a new "relative" model of the universe.

Albert Einstein was born in Ulm, Germany, in 1879. He later became a citizen of Switzerland and the United States. Long before the rise of Hitler, he remarked with gentle irony: "If my [relativity] theory is proved correct, Germany will hail me as a great German, and the French will hail me as a citizen of the world. If it is proved false, the French will call me a German, and the Germans will call me a Jew."[1] Einstein's scientific ideas had tremendous impact, but the Nazis still labeled them "Jewish."

Newton had his apple, moon, and early farm experiences. Einstein had his magnetic compass and geometry book. When Einstein was four or five years old, his father showed him a compass. The youngster was indelibly impressed; there was something consistent and alluring about the direction-pointing needle. This wonder led to the mysteries of magnetic fields. Eventually, Einstein successfully extended Maxwell's electromagnetic field equations to gravitation in a universe of four dimensions. But long before that, at twelve years of age, he became fascinated with Euclid's plane geometry for its "purity, lucidity, and certainty of thinking." The young Einstein, in responding to geometry, already was searching for order. Individuals tend to repeat the history of the race. In geometry, the boy Einstein had developed as far as the Greeks. But gifted individuals come up to date and leap forward from there. Einstein's early interest in plane Euclidean space led to an unorthodox picture of a mildly curved, non-Euclidean universe.

As a "gymnasium" (high school-junior college) student, Einstein was not the most promising. He showed no special interest in or aptitude for class routines; instead he was off on imaginative tangents of his own. Einstein's *Autobiographical Notes* emphasizes that he "soon learned to ferret out that which was able to lead to fundamentals, and to turn aside from everything else, from the multitude of things which clutter up the mind, and direct it from the essential." Reacting against unrelieved rote routines and deadening details, Einstein continues, "It is, in fact, nothing short of a miracle that the modern methods of instruction have not yet entirely strangled the holy curiosity of inquiry; for this delicate little plant, aside from stimulation, stands mainly in need of freedom; without this the plant goes to wreck and ruin without fail." It was probably during a class reverie that Einstein asked himself whether a student running at the speed of light could see himself in a mirror at arm's length in front of him. This little student-and-mirror barb later flowered into the special theory of relativity. Dilemmas like that of Michelson-Morley fell upon prepared and fertile soil. Respect for facts need not mean "cluttering the mind with detail," but rather searching for order through facts.

Einstein obtained a degree in experimental physics at Zurich Polytechnic Institute in 1900 with no special honors, and in 1902 he became a Swiss patent office examiner. However, while at the Institute Einstein had acquired considerable laboratory experience and had spent a great deal of time reading current physical theory. In 1905 he electrified the world with two papers, one on the special theory of relativity and the other on the photon theory of light. A third paper that year on Brownian motion (i.e., the random motion of molecules) led to determination of the number of molecules in a given amount of substance. He also completed a doctoral dissertation based on improved technique in determining the size of molecules. Einstein was then invited to lecture at Bern University. He left the patent office in 1909 to become an assistant professor at Zurich Polytechnic; in 1911 Einstein was made a full professor at the German University in Prague, and in 1913 professor of theoretical physics at the Kaiser Wilhelm Institute in Berlin, then one of the world's leading research centers. While there he published his general relativity theory (1916). In 1921 Einstein received the Nobel prize in physics. Despite his world renown, however, he was forced out of the Wilhelm Institute in 1933 by the Nazis. At that time Einstein accepted an invitation to join the Institute for Advanced Studies at Princeton, where he remained for the rest of his life.

Although he became famous, Einstein's personal life, dress, and manner were always exceptionally simple. As expressed in his autobiography, his "major interest disengaged itself to a far-reaching degree from the momentary and the merely personal, and turned toward the striving for a mental grasp of things." Striving for personal prestige and comfort was alien to him. Incorruptible, he was one of those profoundly simple men who create as they search for order.

Einstein's last great effort was a unified field theory in which he successfully reduced electromagnetic and gravitational fields to a single set of formulas. Time is still needed to determine whether the universe will support that supreme field synthesis. Some time before he died on April 18, 1955, the huge Riverside Church in New York included a statue of Einstein as the only living person among its famous collection of the world's greatest scholars.

Every theory rests on assumptions. Einstein developed his special theory of relativity from the following three main assumptions:

1. *The velocities of all bodies are relative.* Newton himself had acknowledged this relativity principle. Let us say that two trains leave a station together at 40 mph. With respect to each other, the trains are not moving. Only by noticing other outside objects (trees, grass, houses) can the passengers be aware of any motion. And that motion, the 40 mph, is *relative* to these observed objects which themselves are moving. The earth bearing them is moving at an unknown absolute velocity. The earth revolves with a speed of 19 miles a second *relative* to the sun, which revolves with a speed *relative* to the center of the Milky Way, which moves *relatively* to other moving galaxies moving *relatively* to one another. There is no fixed framework in space. Absolute velocity is a meaningless concept.

Again, car *A* speeds at 80 mph relative to the earth. Car *A*, however, moves 20 mph relative to car *B*. which is traveling at 60 mph in the same direction. But the earth also moves. The absolute velocity of the earth and, therefore, of the cars cannot be found. But if an ether exists, it is undetectable, said Einstein, and absolute velocities are unknowable. The situation is circular. We had better forget the ether and absolute velocities.

2. *The speed of light is an absolute for all observers at constant velocities.* The velocities of material bodies are relative, but the speed of light is absolute. In Einstein's words, "The velocity of light in empty space always has its standard value [*c*], independent of the motion of the source or receiver of light."[2] The speed of light would be measured as 186,000 miles a second[3] by an astronomer on earth as well as by one the ray overtakes on a spaceship moving 100,000 miles a second away from the earth. The 100,000 miles a second of the second observer simply would not count. In this postulate, Einstein accepts the

Michelson-Morley results as a fact of nature to be used as a starting point for a new system of ideas.[4]

3. *Laws of nature are the same in all systems moving at constant velocity relative to one another.* This assumption expresses a basic confidence in a unity of nature. It is not just mechanical laws but *all* laws of nature that are the same for inertial systems—that is, for systems not changing velocities. Since absolute velocity is unknowable, the constant velocities of systems are expressed as "relative to one another."

All three assumptions involve observers in nonaccelerating, *inertial frames of reference.* The observers move at constant velocity, relative to one another. In his special theory of relativity (1905), Einstein confines himself to natural laws of inertial reference frames. The earth may be considered an inertial system; its acceleration is negligible. In his general theory of relativity (1916), Einstein extends his attention to *accelerating reference frames* (chapter 17).

Now Is Not Now Everywhere

For ages men assumed that *now* is now everywhere. We see Arcturus in the sky and record its position at this "instant." Actually, the light has traveled thirty-eight years to reach us. Meanwhile, Arcturus has moved on and actually may no longer exist. An observer midway between us and Arcturus would have seen that star nineteen years before we did. The same instant of Arcturus's given position is separated by nineteen years for the two observers. *Now* is therefore not absolute, not the same everywhere. In assuming an absolute now, or simultaneity, in the universe, men had made no allowance for the speed of light. They had treated light as if it were infinite in speed, whether from Arcturus or anywhere. Einstein concluded that "Every reference body (or coordinate system) has its own particular time; unless we are told the reference body to which the statement of time refers, there is no meaning in a statement of the time of an event."

But there is a bigger problem than that of a different "now" at two different places. It is the problem of two events occurring together as seen from one place, and not together when seen from another place. Two things happening for us here *now* may not be happening together for someone elsewhere. Einstein posed this problem of simultaneity in the following thought experiment.

Two bolts of lightning strike a railroad track at points *A* and *B* (Fig. 16.2). The bolts are simultaneous to an observer *O* situated along the track exactly midway between points *A* and *B*. A fast train moves from *B* to *A* with a passenger *P* directly opposite observer *O* at the time of the flashes. In a mirror arrangement, *P* sees the flash at *A* slightly sooner than the flash at *B* because of his motion toward *A*. In fact, if the train were moving as fast as light, *P* would see only the flash at *A*. The light from the flash at *B* would not catch up to observer *P*. Flashes simultaneous to observer *0* would, therefore, not be simultaneous to *P*.

Einstein's thought experiment led to his famous relativity principle of simultaneity that events occurring together for one person may not be together for another due to the relative motion of the two observers. But if events occur together for one person and not for another, how could time be the same for both? If a lapse of time for one does not exist for the other, how could a a second for one be a second for another?

SPACE AND TIME AS SHADOWS

The absolute velocity of light is a cornerstone of Einstein's system. No matter how fast an observer moves, he will always find that light travels 186,000 miles a second with respect to him. The observer may be on earth or on a spaceship leaving the earth at 100,000 miles a second; light still has a speed of 186,000 miles a second with respect to him. It is as if the extra 100,000 mile-a-second velocity of the spaceship does not exist. How is this possible? Involved is a contradiction between an absolute speed of light and an ordinary addition of velocities.

It was in facing this contradiction that Einstein threw out space and time as absolute entities. Since the observed absolute speed of light does not yield, the space and the time intervals that went to make up the light measurements must. If no matter how fast an observer moves, light always passes by him at 186,000 miles a second, then perhaps a mile and a second as measured from a spaceship are not the same as they are on earth. Perhaps Newton was incorrect in his assumption that "the duration of an event is independent of the state of motion of the system of reference." Perhaps space and time measurements do depend upon the velocity of an observer. That is, a space or time measurement is a measurement peculiar to a given observer; therefore, value valid for one may not be so for another. Determining the velocity of light in miles per second involves measurement of space and time as well as a problem of simultaneity. And simultaneity is relative. Where does all this lead? Clearly, Einstein was faced with the problem of how to correlate clocks and metersticks in one system with those in another. Confident of the possibility of such correlation, he expressed his basic postulate, "All laws of nature are the same for all uniformly moving systems." The Newtonian concept of relativity referred to *mechanical* laws being the same for all uniformly moving systems. To mechanical law, Einstein pointedly added all the other laws of nature, such as

Figure 16.2

Relativity of simultaneous events. For observer O, the lightning flashes at A and B occur together. For passenger P in the moving train, flash A occurs before flash B.

laws of light, electromagnetism, or heat. If there is no absolute framework of space or time for moving bodies, there is none for light or other radiation. Consequently, there is no detectable ether. Einstein was also indicating that in a universe where everything is in motion relative to observers also in motion, natural laws exist for correlating all measurements from one system of motion to that of another. Space and time, no longer fixed frameworks, became merely forms of perception or ways of measuring that give different results in different systems. The laws that Einstein sought were relationships between space, time, and other measurements from different systems moving uniformly relative to one another.

But how were these laws or relationships to be found? With space, time, and simultaneity varying from one moving system to another, how could observations of events in one moving system be specifically linked with observations of the same events in a differently moving system? Einstein's clue was the constant velocity of light that began the difficulty. That all observers regardless of motion obtain the same value for the speed of light was an empirical fact. The speed of light c replaced space and time as an absolute; the constant c, therefore, became the key to an orderly universe. If space and time varied from one system to another, the problem narrowed down to these measurements so varying that no matter how relatively fast or slow the observer was going, he always found light to pass by him at the same velocity, 186,000 miles a second. Mile and second measurements must change accordingly. Unknown to an observer, his clocks must slow down as the velocity of his reference system increases, and his measuring rods must shrink in the direction of motion.

In a sense, Einstein was starting from a set answer ($c = c$ for all) at the back of the book and was working his way to unknown variables of space, time, or mass at the front. In this way, Einstein hoped to link measurements varying from one reference frame to another. And if in the process, the answer at the back of the book became a postulate for a new conceptual model of special relativity, then remember that the new postulate was based upon such facts of nature as the Michelson-Morley results.

Einstein found what he wanted in the Fitzgerald-Lorentz transformation equation, particularly in the contraction factor

$$\sqrt{1-\frac{v^2}{c^2}}.$$

But he found it with v as the relative velocity between an object and an observer, and c as the constant speed of light with respect to the observer. That is, v and c are determined with *respect to an observer and not to an undetectable fixed ether.*

By use of the above contraction factor, Einstein found that he could predict the shrinkage of an object's length, a slowing down of clocks, and an increase of an object's mass according to the relative velocity of an observer. Let us now illustrate.

SPACE CONTRACTS, RULERS SHORTEN

Spaceship S returns to the earth at a velocity v of 161,000 miles a second. The velocity of S is, of course, relative to the earth. An observer in S finds that a light ray passes him at 186,000 mile-a-second relative to himself. An earthly observer E finds that the same ray passes him at the *same speed* relative to himself. It was as if the 161,000 mile-a-second speed of the spaceship back to the earth did not exist. For the speed of the light ray to be the same in both places, miles and seconds in them cannot be the same. There must be a relative shrinkage of metersticks and slowing of clocks in one of the two places for the speed of light to come out the same for both. The transformation equation that predicts the appropriate shrinkage of *measuring* rods or objects in a frame of reference is as follows:

$$l = l_0 \sqrt{1-\frac{v^2}{c^2}},$$

where l_0 = the length of an object at rest to an observer,
  l = the length of the object in relative motion to an observer,
  v = the relative velocity between object and observer, and
  c = the speed of light.

Suppose observer S measures the length l_0 of his spaceship and finds it to be 100 feet. What length l would an observer on earth obtain for the spaceship returning to earth at a velocity v of 161,000 mile a second? With the speed of light c being 186,000 miles a second, the above transformation equation gives an answer of 50 feet. That is,

$$l = l_0 \sqrt{1 - \frac{v^2}{c^2}}$$

$$= 100 \sqrt{\frac{1 - (161,000)^2}{(186,000)^2}} = 100 \sqrt{1 - .75}$$

$$l = 50 \text{ ft.}$$

The spaceship is 100 feet long for the pilot and 50 feet long for observer E. The spaceship pilot is unaware of the contraction of S because his yardstick also shrank to one half (as observed from the earth). Thus, we see two different measurements, l and l_0, of the length of the same object by observers in two different systems. But

$$\sqrt{1 - \frac{v^2}{c^2}},$$

the relativity factor, permits correlation between the two systems regardless of relative speed. The universe is still one; laws of nature are universal.

Space contraction works both ways. Both observers S and E find that objects flatten in the other frames of reference. The space pilot finds the earth's diameter flattened in the direction of motion from 8,000 miles to 4,000 miles. In the above equation, l_0 is the 8,000-mile diameter determined on earth, and l is the 4,000 miles from the spaceship. The ratio $l_0{:}l$ remains 2:1. The l_0 always represents the length of an object at rest with respect to an observer.

Contraction of an object is in the direction of its motion *only;* that is, its length. There is no contraction in the other two dimensions, width and height. They remain the same for all observers.

When the spaceship is back on earth, the length measurements by observers S and E are the same, 100 feet. That is, with the spaceship velocity v now zero for observer E:

$$l = l_0 \sqrt{1 - \frac{v^2}{c^2}} = l_0 \sqrt{1 - \frac{0}{c^2}} = l_0 (1) = 100 \text{ ft.}$$

The situation is now Newtonian and illustrates that when the relative velocity v between an observer and a thing observed is zero or small compared to the speed of light, the Newtonian and Einsteinian answers will be the same or close. Newtonian mechanics is a special case (low relative velocity) of Einstein's relativity.

Now suppose the spaceship could travel away from the earth with the speed of light. The relative velocity v between the spaceship and the earth would equal c. Therefore,

$$l = l_0 \ \sqrt{1 - \frac{v^2}{c^2}} = 100 \ \sqrt{1 - \frac{c^2}{c^2}} = 100 \ \sqrt{1 - 1} = 100 \ (0) = 0.$$

As seen from the earth, the length of the spaceship would have contracted to zero—an impossibility! That is, material objects may closely approach but never reach the speed of light in a vacuum. Light has the top speed in the universe.

TIME EXPANDS, CLOCKS SLOW DOWN

Time runs differently for observers moving at constant speed with respect to each other. How does this idea follow from Einstein's postulate that observers at different speeds obtain the same value in measuring the speed of light? In answer, Einstein provides the following thought experiment (Fig. 16.3 A–C).

(A) Train passengers P_2 and P_1 are in line with observer O when P_2 strikes a match. (B) Because the train is moving, the match light takes a longer path d_0 in time t_0 for observer O, as compared to the path d_1 in time t_1 for passenger P_1.(C) If c is the speed of light on path d_0 and v is the velocity of the train, the arm of the right triangle is $\sqrt{c^2 - v^2}$. (The square of the hypotenuse of a right triangle equals the sum of the squares of the two sides.) By similar right triangles (B and C),

$$\frac{d_1}{d_0} = \frac{c^2 - v^2}{c^2}.$$

Since the speed of light is the same ($c_1 = c_0$) relative to all observers,

$$\frac{d_1}{t_1} = \frac{d_0}{t_0}.$$

(Velocity c is distance d divided by time t). By proper algebraic substitution,

$$t = t_0 \ \sqrt{1 - \frac{v^2}{c^2}}$$

where t_0 = the time interval of an event *inside* an observer's
frame of reference,

t = the time interval of the same event measured by an
outside observer,

v = the relative velocity of the two reference frames,
and c = the speed of light.

The above equation is the famous Einstein-Lorentz time transformation equation as derived by the Pythagorean theorem. Clocks in different reference frames keep different time in accordance with the motion of the observer. The transformation equation links the different clocks together when the motion is uniform.

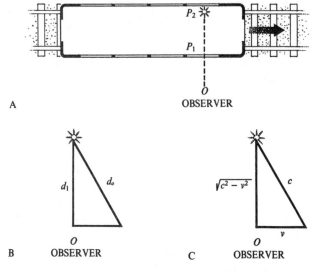

Figure 16.3

Time measurement as interaction. (A) Passengers P_2 and P_1 are in line with observer O when P_2 strikes a match. (B) Because the train is moving, the match light takes a path d_0 for observer O, as compared to the path d_1 for passenger P_1. (C) If c is the speed of light on path d_0 and v is the velocity of the train, the arm of the right triangle is $\sqrt{c^2 - v^2}$. (The Square of the hypotenuse of a right triangle equals the sum of the sqaures of the two sides.)

In a further illustration, suppose that identical clocks on spaceship S and on earth are synchronized when together. When S moves away from the earth with a relative velocity v, observer E on earth and S on the spaceship find that the clocks in the two frames of reference do not keep time together. At a relative velocity of 161,000 miles a second, E finds that S's clock runs half as fast as his. This may be shown by inserting the values for v and c in the time equation and solving for the ratio t/t_0. Paradoxically, each observer finds that the clock of the other slows down.

At a relative velocity of 93,000 miles a second, each observer finds that the clock of the other runs about nine-tenths as fast as his. If the spaceship should reach the speed of light ($v = c$), time would stop. An upper limit in speed would be reached. At the other extreme, when the spaceship is at rest on earth and its velocity $v = 0$, its clock keeps time together with the clock on earth. That is, the time equation then reduces to $t = t_0$. The situation now is Newtonian; time is the same in both frames of reference.

Let us now add to Chaucer's previously quoted lines about time in order to illustrate its relativistic motion:

> For though we sleep or wake, or roam or ride,
> Ay flees the time, it will no man abide.
> It now depends on how fast we ride.

There are no equations here, but the third line can be interpreted as relativistic for the fourteenth century.

ONE PLUS ONE EQUALS TWO—SOMETIMES

If two spaceships pass the earth in opposite directions at 100,000 miles a second, by ordinary arithmetic, they should pass each other at 200,000 miles a second. But that is impossible! By the special theory of relativity, no relative speed can exceed that of light, which is 186,000 miles a second. The question is, how can the velocities of any two objects be added so as never to exceed the speed of light? The following equation is Einstein's relativistic answer:

$$v_3 = \frac{v_1 + v_2}{1 + \dfrac{v_1 v_2}{c^2}}$$

where $v_3 =$ the relative velocity of two objects to each other
(e.g. two spaceships);

$v_1 =$ the relative velocity of one object to a third
(e.g., to the earth);

and $v_2 =$ the relative velocity of the second object to a third
(e.g., the earth).

By placing the 100,000 mile-a-second velocity value of each spaceship in Einstein's addition-of-velocities equation, we obtain a total of only 155,000 miles a second, shown as follows:

$$v_3 = \frac{v_1 + v_2}{1 + \dfrac{v_1 v_2}{c^2}} = \frac{100,000 + 100,000}{1 + \dfrac{(100,000)\,(100,000)}{(186,000)^2}} = 155,000 \text{ mi/sec.}$$

Although the spaceships pass each other in opposite directions at 100,000 miles a second with respect to the earth, their velocity relative to each other is 155,000 miles a second, not 200,000 miles a second. And 155,000 miles a second is lower than the speed of light.

Now suppose we take the extreme case and raise the speeds v_1, and v_2 of the spaceships each to the speed of light c. That is, $v_1 = c$ and $v_2 = c$ in respect to the earth. With Newtonian addition of velocities, spaceships 1 and 2 would travel at twice the speed of light ($2c$) relative to each other since they travel in opposite directions.

By Einstein's equation the total value is still only c:

$$v_3 = \frac{v_1 + v_2}{1 + \frac{v_1 v_2}{c^2}} \ = \ \frac{c + c}{1 + \frac{(c)\,(c)}{c^2}} \ = \frac{2c}{1 + 1} = c.$$

That is, although both spaceships are traveling in opposite directions at the speed of light relative to the earth, they are traveling only at the speed of light with respect to each other (Fig. 16.4).

At low speeds the Newtonian addition of velocities closely approximates the Einsteinian. Suppose two cars zoom past each other at 100 mph relative to a tree. They flash by at 200 mph relative to each other whether measured according to Newton or Einstein. This may be checked by using the Newtonian and Einsteinian equations above.

Once again we find that Newtonian principles are special cases of the Einsteinian. Newton and Einstein begin to merge for objects at relative speeds below one-tenth that of light. Under the ordinary speed of a train, plane, or rocket, the slowing down of a watch or shrinkage of a ruler is too minute to matter. It is when a relative speed begins to approach that of light that speed really counts. Einstein did not overthrow Newton. Newton's model retains a place in the larger picture of relativity.

YOUR WEIGHT INCREASES WITH SPEED

In 1901 Kaufmann and J. J. Thomson discovered that electrons increase in mass with speed. This fact was another challenge to Einstein. Newton's laws certainly did not provide for such changes in an object's mass. Weight is a gravitational force upon an object that changes with distance from the earth; but an object's mass depends upon the "quantity of matter" within it, which supposedly does not change. If, however, an electron's mass did increase with speed, so would the mass of anything else. Is an object's mass, like its length, a measurement relative to its speed?

In answer, Einstein proposed that every object has a basic *rest mass* m_0 when at rest relative to an observer. When moving, body acquires extra mass Δm. A moving object at any time has total mass m equal to its rest mass m_0 plus an extra mass Δm from its motion. That is, $m = m_0 + \Delta m$. The faster an object moves relative to an observer, the more extra mass Δm it acquires and the more total mass m it has.

But just how much does velocity change the mass of an object? Einstein found that the same relativity factor applied to mass, as to length or time, only inversely—that is, in the form $1\sqrt{1 - v^2/c^2}$. If m_0 is the rest mass of an object, its total mass m measured when moving at velocity v relative to an observer is

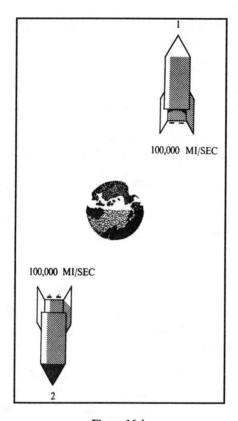

100,000 MI/SEC

100,000 MI/SEC

Figure 16.4

Adding velocities in relativity. Nothing travels faster than the speed of light (186,000 mi/sec). If two spaceships pass the earth in opposite directions at 100,000 mi/sec, by ordinary arithmetic, they should pass each other at 200,000 mi/sec. But by relativity theory, the spaceships have a velocity of only 155,000 mi/sec with respect to each other.

$$m = \sqrt{\dfrac{m_0}{1 - \dfrac{v^2}{c^2}.}}$$

From this equation, a given object increases in mass with speed relative to a given observer. From the same equation, two or more observers at different speeds obtain different values for the total mass *m* of that same object. Similar differences in measured lengths and travel times of objects for various observers were noted in earlier sections. To illustrate, spaceship *S* has a 1,000-ton rest mass (m_0). *S* moves away from the earth at 161,000 miles a second and passes spaceship X at a relative speed of 93,000 mile a second. By the above relativistic mass equation, spaceship S at these relative speeds has a mass of 2,000 tons as measured from the earth, but only 1,200 tons as determined from spaceship X that it passes at a lower relative speed. From this it can be seen that the faster a spaceship travels relative to an observer, the greater is its mass for that observer. At the speed of light c, the mass of a spaceship would be limitless.

An object's weight increases along with its mass. An increase in speed, therefore, increases weight as well as mass.

Again, the speed of light is seen as a top speed in the universe. No amount of accelerating force could make an "infinite" mass move faster. Electrons, however, have been sent swirling at over 0.999999999 the speed of light c, to attain over 40,000 times their mass at rest. For objects at low speeds, the increase in mass is too small to be easily detected. Masses may then be treated as constant, and Newton's laws applied.

THE AMAZING $E = MC^2$

But Einstein did not drop relativity of mass at this point. From the idea that objects increase in mass as they go faster, he drew the amazing conclusion that matter and energy are equivalent. After all, mass m did not increase alone; kinetic energy, or energy of motion ($\frac{1}{2}mv^2$), also increased with velocity v. Perhaps, said Einstein, matter and energy are different aspects of the same thing: matter could be a concentrated form of energy. Energy has been distinguished from matter by having no apparent weight. Perhaps energy in free form as light, heat, or kinetic energy has weight, but not enough for detection. Matter as concentrated energy would show weight. In that case, said Einstein, matter has fantastic amounts of concealed energy to be unlocked. How fantastic when first proposed and yet shockingly real in today's atomic era! Science can indeed be stranger than fiction.

Einstein equated free energy with matter through the following equation, one of the most significant in history:

$$E = mc^2$$
where E = the free energy in ergs,*
m = the mass in grams, and
c^2 = the speed of light squared (900 billion billion cm^2sec^2)!

---

*An erg is the metric unit of work in which the work done is by a force of one dyne acting through a distance of one centimeter.

A small amount of mass *m* as concentrated energy must be multiplied by c^2 to obtain an equivalent in unbound, or free, energy *E*. The amount of free energy is tremendous because of the c^2.

An iron ball should accordingly weigh more when red hot than when cool. Energy has mass (and mass, energy). The heat energy should give the ball extra mass Δm. Radiation is energy, and should have mass. The sun, stars, or a heated iron ball should lose mass as they radiate.

A one-pound lump of coal completely converted into free energy would equal all the electric power output in the United States for about a month. Conversely, the amount of heat needed to change 30,000 tons of water into steam would weigh only one-thirtieth of an ounce. Very seldom, however, is a total mass *m* completely converted into its total energy *E*. The sun, stars, or H bombs lose small portions of their mass Δm as radiation ΔE. In such cases, $E = \Delta mc^2$. The $E = mc^2$ equation represents total matter-energy transformations, and the ΔE equation partial matter-energy transformations.

The law of conservation of energy claims that the total energy of the universe is constant, that energy can be neither created nor destroyed. Does not Einstein's $E = mc^2$ contradict this basic law? When matter changes to free energy, is not energy added to the world? The answer is no. Einstein changed the concepts of matter and energy. Matter is a special form of energy, and all matter is included in the energy total. The law of conservation of energy becomes the law of conservation of matter-energy stating that the total amount of both matter and energy in the universe (or any closed system) is always the same.

SEEING IS BELIEVING

It is one thing to build a mathematical model like relativity, and another for the model to hold in the world of things. Will the universe support ideas of shrinking rods, slowing clocks, and increasing masses? What is the evidence?

Kaufman (in 1901 and 1906) and E. Bucherer (in 1908) experimented with fast-moving electrons, to find that the higher the speed *v* of an electron, the greater its mass. When, after 1905, they applied Einstein's relativistic mass equation to their data, they found that all electrons had the same rest mass m_0. That is, when the different values of *m* and *v* were appropriately inserted in the equation, m_0 came out the same for all electrons. With greater speed, each electron had increased in mass from a common rest mass as predicted by relativity. Here was Einstein's first evidence. A number of years later, Professor E. O. Lawrence of the University of California found it necessary in the development of his cyclotron to make relativity corrections for the increased mass of his accelerated particles. Electron mass has been measured under various circumstances. The particular speed of electrons in television "picture" tubes gives them an approximate mass increase of 5 percent. Electrons whirled in accelerators at over 0.999999999 the speed of light c have attained 40,000 times their mass at rest, as expected. At the speed of light, the electrons, of course, would have infinite mass. Protons, the nuclei of hydrogen atoms, also have been accelerated to speeds approaching that of light, with the expected increase of mass.

Comets form tails as they swerve around the sun due to pressure of the sun's light. This pressure suggests that radiation has mass. Quantities of metal heated to high temperatures have definitely shown slight increases in mass in accordance with $\Delta E = \Delta mc^2$. Heat energy thus shows a mass equivalent. Atom-smashing apparatus, atomic power plants, and atomic bombs are very much a part of our world: their processes provide everyday evidence for Einstein's idea of the interconversion of matter and energy. For better or worse, Einstein's $\Delta E = \Delta mc^2$ has been successful. Social intelligence can make it far better.

The relativity of time has also been verified. In 1936 H. E. Ives of Bell Telephone compared hydrogen atoms at rest to others at high velocity. He found a difference in the vibration frequency of the atoms exactly as predicted by relativity. A change in atom vibration frequency is analogous to a change in heartbeat rhythm or in the ticking of a clock. Atomic particles called *mesons,* with very short half-lives, were found in experiments conducted in 1941 to have a slower rate of aging as they were accelerated to ever higher velocities. In 1956 Professor Frank S. Crawford at the University of California found that the half-lives of some subatomic particles moving through the atmosphere toward earth at high velocities are fifteen times as great as those of identical particles at rest.

The slowing down of time is reciprocal: To observers S and E moving at 161,000 miles a second relative to each other, time appears slower in the other's spaceship. The heart beats away time, too; all physiological processes are time processes. To S, E should appear to age more slowly than himself; to E, S should appear to age more slowly. Each should find that the other's physiological processes slowed down. Let us say that S and E were identical twins thirty years of age when S left the earth. When S returned after twenty years, S and E would each expect the other to appear younger than himself by ten years. How could both be right? This contradiction, the famous clock paradox, seems to disprove relativity. Actually, it does not: The spaceship accelerates to a high speed on leaving the earth, decelerates and accelerates in reversing direction at a furthermost point, and decelerates again in returning to the earth. Any curved or circular motion also involves acceleration. The special theory of relativity does not cover these periods of acceleration; Einstein's later general theory of relativity does. The paradox, therefore, is no paradox. While the spaceship accelerates, the earth remains an inertial frame. Time effects are reciprocal only when both reference frames are at constant relative velocity. During periods of acceleration, observer *S* ages less than observer *E* back home, and *S* will therefore return home appearing younger to all, including himself.

In 1958 the Soviet scientist P. A. Cerenkov shared the Nobel prize in physics with his colleagues Ilya M. Frank and Igor Y. Janns for his discovery of Cerenkov radiation. The speed of light in water is only three-fourths its speed in a vacuum, and in glass only two thirds. Electrons approach the speed of light in a vacuum. Can electrons travel faster than light in water, air, or another medium, where the speed of light drops considerably? Cerenkov successfully sent electrons through water at a speed faster than light in that medium. When this occurred, a cone of light waves, unique in color, showed itself very much like the pattern of bow waves which form when a ship moves faster than water waves. The Cerenkov effect does not contradict Einstein's conclusion that light and other electromagnetic waves are the fastest things in the universe. Electrons or other particles may travel faster than light in material media, but Einstein's reference is to a vacuum.

CONCLUSION

Einstein examined classical mechanics at its foundations. No system of ideas is stronger than the assumptions upon which it rests. Could Newtonian assumptions of space, time, mass, energy, force, field, or gravitation be creating difficulties? In examining bedrock concepts and assumptions, Einstein thought his way through to his special (1905) and general (1916) theories of relativity, with their resulting revolutionary impact on modern physics, life, and thought. His relativity model solved problems that previous models could not. Relativity ushered in atomic energy. It provided theoretical foundations, and Newtonian mechanics was absorbed as part of a larger picture. Great engineering feats have been achieved with Newtonian principles based on standard clocks and metersticks. Under ordinary speeds of autos, planes, or rockets, the slowing down of a watch or the shrinking of a ruler is too infinitesimal to matter. But when atomic particles approach the speed of light, Newtonian principles fall short; and when matter transforms into energy, they fall flat. The physical universe is only partly Newtonian. We now must be concerned about relative motion between an observer and what he observes. What is obtained in *measurement* varies according to this relative motion relationship. Predictably so.

We must now distinguish between (1) actual events, (2) their signals (light and other radiations), (3) our observations, and (4) our interpretations. Einstein has added to our scientific and philosophic maturity.

NOTES

1. Albert Einstein, *Out of My Later Years* (New York: Philosophical Library, 1950).
2. Albert Einstein, *Relativity,* trans. Robert W. Lawson (New York: Crown, 1961).
3. A closer approximation than 186,000 mi/sec is 186,380 mi/sec but for our purposes here, the former is close enough.
4. Operationalism had its philosophic origins in the position of Ernst Mach's revolutionary *Science of Mechanics,* 1883. The germ idea was that any scientific idea, as an ether supposedly permeating space, should not be used unless the idea could be put to a test. In respect to the concept of ether, the Michelson-Morley experiment showed that the existence of an ether could not be determined. Einstein took the cue: He neither used nor needed the ether concept in developing his special theory of relativity. Nor did he use the ether concept in developing a four-dimensional space-time model of gravitation in his general theory of relativity, as will he described in Chapter 17.

SUGGESTIONS FOR FURTHER READING

Clark, Ronald. *Einstein, His Life and Times.* New York: Avon, 1972.

Coleman, James A. *Relativity for the Layman.* New York: Signet, 1962.

D'Abro, A. *Evolution of Scientific Thought—From Newton to Einstein.* New York: Dover, 1950, parts I and II.

Davies, Paul. *Superforce, Search for a Grand Unified Theory of Nature.* New York: Touchtone, 1985.

De Broglie, Louis. *Physics and Microphysics.* New York: Pantheon, 1966.

Durell, Clement. *Readable Relativity.* New York: Harper, 1960. chs. 1–7.

Eddington, Arthur. *Space, Time and Gravitation.* New York: Harper, 1959.

Einstein, Albert. *Out of My Later Years.* New York: Philosophical Library, 1950.

———. *Relativity.* New York: Crown, 1952. Part 1: "Special Theory."

Einstein, Albert, and Leopold Infeld. *The Evolution of Physics.* New York: Simon & Schuster, 1952.

Frank, Philip. *Relativity: A Richer Truth.* Boston: Beacon, 1950.

Hawking, Stephen W. *A Brief History of Time.* New York: Bantam, 1988.

Infeld, Leopold. *Albert Einstein: His Work and Influence.* New York: Scribner, 1956.

Jaffe, Bernard. *Michelson and the Speed of Light.* Garden City, N.Y.: Anchor, 1960.

Kaufmann, William J., III. *Black Holes and Warped Space-Time.* New York: Bantam Books, 1979.

Landau, L. D., and G. S. Rumer. *What Is Relativity?* New York: Basic Books, 1960.

Reichenback, Hans. *From Copernicus to Einstein.* New York: Philosophical Library, 1942.

Rucker, Rudolf. *Geometry, Relativity and the Fourth Dimension.* New York: Dover, 1977.

Russell, Bertrand. *The ABC of Relativity.* New York: Mentor, 1959.

Schilpp, Paul Arthur. *Albert Einstein, Philosopher-Scientist.* New York: Tudor, 1951.

Will, Clifford M. *Was Einstein Right? General Relativity Tested.* New York: Basra Books, 1986.

17

The Strange World of Four-Dimensional Space-Time: Einstein's Gravitational Framework

The universe is not a rigid and inimitable edifice where independent matter is housed in independent space and time; it is an amorphous continuum without any fixed architecture, plastic and variable, constantly subject to change and distortion. Wherever there is matter and motion, the continuum is disturbed. Just as a fish swimming in the sea agitates the water around it, so a star, a comet, or a galaxy distorts the geometry of the space-time through which it moves.

Lincoln Barnett, *Einstein*, 1948

The last chapter focused on the special theory of relativity, in which concepts and laws of physics are developed for those situations where reference frames move at *constant velocities* relative to each other. We now turn over attention to those physical situations in which observers and what they observe *accelerate* in respect to each other.

LOOKING INTO SPACE-TIME

The great German mathematician Hermann Minkowski (1864–1909) was exceedingly impressed by the relativity of space and time. This former teacher of Einstein opened a science meeting in Cologne in 1908 with these words: "The views of space and time which I wish to lay before you have sprung from the soil of experimental physics, and therein lies their strength. They are radical. Henceforth space by itself and time by itself are doomed to fade into mere shadows, and only a kind of *union* of the two retains an independent reality" (italics added). Minkowski proposed a new concept; space-time. He combined the three dimensions of "shadowy" space and an independent dimension of time into a four-dimensional space-time.

There is nothing mystical about compounding space and time, whether in mathematical models of the universe or in everyday life. Many simple examples show time hidden in the fabric of space, or better still, show space and time as part of the same space-time fabric of events. Solar system motions and distances determine calendars and clocks. The earth completes an orbit to mark off a year. Other planets with shorter or longer orbits (distances) have shorter or longer years (time). At noon, the sun is more directly overhead; a point in time is set by the sun's relative position in space. A day is determined by the rotation of a point on the earth through a full circle. A day on the moon is over twenty-seven times longer than ours; a point on the moon's surface completes a different circle at a different rate. The complete swing of a pendulum (time) varies with a changing acceleration of gravity at different places (space) on earth. Seasons (time) are due to an angle of inclination of the earth's axis as the earth moves in its orbit.

Basically, when we look into the heavens, we look into space-time. If we say that a star is 200 light-years away, we are indicating not only distance but time. The light now entering our eyes left the star 200 years ago and indicates the star's position *at that time.* Meanwhile the star has moved on. The only proper frame of reference for the physical reality of that star is a combined space-time reference, a "four-dimensional space-time continuum." Space or distance has meaning or reality only in terms of time, of when.

A train schedule provides a good example of the necessity for describing position in terms of time. We are interested not merely in the route of a train, but *where* it will be *at what time.* In that sense, the train schedule is a "two-dimensional space-time continuum," with the railroad track just a one-dimensional space component. The distances can be plotted against the time.

On the same basis, a sea captain is concerned with his two-dimensional space continuum of latitude and longitude against time, thus involving a "three-dimensional space-time continuum." An air pilot has a four-dimensional space-time situation. He is concerned not only with his three-dimensional space continuum of latitude, longitude, and altitude as such, but with these in the four-dimensional where-when of his schedule.

Einstein was greatly impressed by Minkowski's concept of space-time. The idea provided insights into the world of events and had promise for the extension of Einstein's relativity theory. The space-time concept is made graphic by plotting space against time. Every event may have a point in a space-time graph. Suppose a rock is dropped form a 400-foot cliff. Figure 17.1A shows the rock's fall in space and in time; Fig. 17.1B, the rock's fall in space-time. Point *A* (Fig. 17.1B) is the position of the rock 400 feet above the ground at time 0; point *B,* 16 feet lower, 1 second later; point *C,* 64 feet from the top, 2 seconds later, etc. These points and any points between them are called *point-events* in space-time. A single point represents the place of an event in a combined space-time graph or coordinate system.

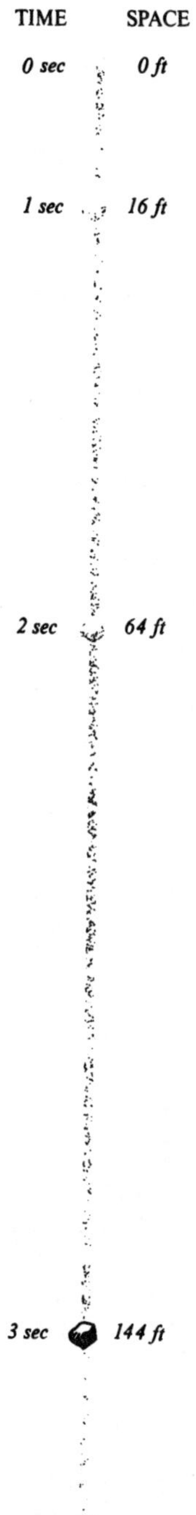

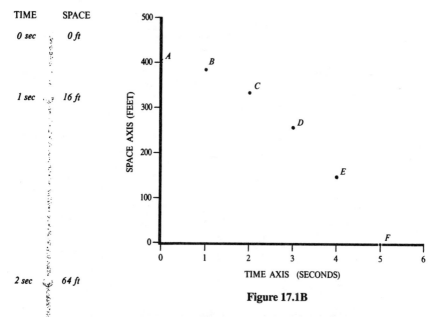

TIME   SPACE

0 sec   0 ft

1 sec   16 ft

2 sec   64 ft

3 sec   144 ft

Figure 17.1B

A rock's fall in space-time. Points *A* through *F* are point-events of this fall. (For comparison with Fig. 17A, only the space axis here is vertical. Frequently in a space-time graph, the time axis is vertical.)

Figure 17.1A

A rock's fall in space and time.

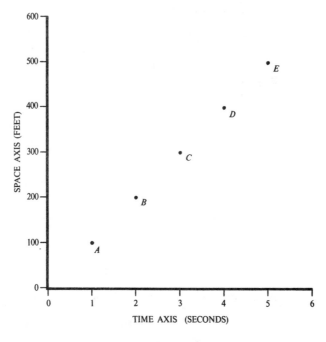

Figure 17.2

The world line in space-time of a car traveling at 100 feet per second for 5 seconds. Points *A* through *E* are point-events in this world line. (The diagonal may be extended in its given direction for as long as the velocity is constant.)

A succession of point-events in a curve is called a *world line.* Curved *AF* is the world line in space-time of the rock during its fall. In this case, one of uniform acceleration of gravity, the world line is a parabolic curve. If a rock fell or moved at constant velocity, its world line would be a straight line (Fig. 17.2). Even a parked car would show a horizontal line on a space-time graph: its "fixed" position has duration in time. Particularly significant for Einstein was the fact that acceleration could be identified with curved lines in space-time. In the development of general relativity, he interested himself in gravity and acceleration.

MAKING RELATIVITY ABSOLUTE

People incorrectly believe that there is nothing absolute in relativity. Actually, space-time became an absolute, an entity that gave the same values to everybody regardless of circumstances, just as did the speed of light in the special theory of relativity. Einstein showed that observers with different relative values for the space and time of an event obtain the same value for space-time distance in the world line of the event. A simplified explanation of this absolute, called the *Einstein interval,* is centered around Figure 17.3A–D.

1. One (space) dimension: $OB = x$. An object travels a distance (OB) in one dimension x.

2. Two (space) dimension: $OB = \sqrt{x^2 + y^2}$. An object travels a distance OB in two space dimensions x and y. The Pythagorean theorem is involved: The square of the hypotenuse of a right triangle equals the sum of the squares of the two sides. OB is the hypotenuse, and x and y are the two sides.

3. Three (space) dimensions: $OB = \sqrt{x^2 + y^2 + z^2}$. An object travels a distance OB in three space dimensions x, y, and z. The Pythagorean theorem as applied to three dimensions is involved.

4. Four (space-time) dimensions: $OB = \sqrt{x^2 + y^2 + z^2 - c^2t^2}$. An object travels a distance OB in four-dimensional space-time, x, y, z, and ct. The equation here also has the Pythagorean theorem form (except for the minus sign before c^2t^2). The time, or fourth-dimension, factor is in ct, the distance light travels at velocity c in time. (Distance equals velocity multiplied by time.)

OB, in the fourth equation above, is the Einstein interval, the distance between two points on a world line. Each observer of the motion of an object may obtain a different space measurement x, y, or z, and time measurement t. Each inserts different space and time values in the first equation but obtains the same value for *OB* in space-time. Thanks to Minkowski, Einstein's equation was not a mathematical device for an absolute quantity *OB* but a reflection of the world "as a four-dimensional space-time continuum." Relative space and time became local aspects of a more basic, common space-time developed by Einstein into a common, universal frame of reference in a general theory of relativity.

THE STRANGE WORLD OF ACCELERATION

Space and time are flickering "shadows." But we have seen that when systems move at constant velocity relative to each other, the different values for space, time, or mass can be equated. Correspondence between systems exists and nature seems orderly. But these relativity equations do not apply when systems accelerate. What relationships exist that include accelerating systems? That was Einstein's basic problem in general relativity: there could not be one set of laws for uniformly moving reference frames and another for accelerating

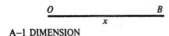

A–1 DIMENSION

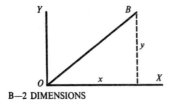

B—2 DIMENSIONS

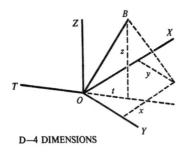

C–3 DIMENSIONS

Figure 17.3A–D

The distance *OB* in terms of 1 to 4 dimensions

A.1 (space) dimension:
$OB = x.$
B.2 (space) dimensions:
$OB = \sqrt{x^2 + y^2}.$
C.3 (space) dimensions:
$OB = \sqrt{x^2 + y^2 + z^2}.$
D.4 (space-time) dimensions:
$OB = \sqrt{x^2 + y^2 + z^2 - c^2 t^2}.$
OB here represents distance in space-time (the Einstein interval).

D—4 DIMENSIONS

systems. Einstein expressed his confidence in the unity of nature in his *general relativity postulate: Laws of nature are the same for all systems regardless of their state of motion.* Einstein, assuming order, proceeded to create order through a general theory of relativity.

Newton's laws of motion emphasize *individual objects* moving at constant velocity (law of inertia) or with acceleration (law of acceleration) *in a fixed space and time.* But Einstein's focus is on *reference frames* of observers *that are inertial or that accelerate.* Newton's laws and special relativity apply to inertial frames of reference and not to accelerating frames. Let us see why.

A stationary disk is an inertial frame of reference. An object can remain at rest on it, and Newton's laws apply to it as on the earth. But the same disk when rotating is an *accelerating* frame of reference. Objects tend not to remain at rest on it but to slide off. An observer on it feels a strange force tugging at him. A ball rolled from the center curves away from the radius. Objects on a rotating disk do not move freely in a Newtonian sense of inertia; Newton's laws no longer apply. In accelerating frames of reference, inertia becomes a force commonly felt in trains rounding curves or buses sharply stopping or restarting. The earth is an inertial frame. Its very slow rotation once a day and revolution once a year may be ignored; objects can remain at rest or at constant velocity on it, and Newton's laws apply for objects at speeds considerably lower than that of light.

Briefly, we know of no frame of reference that is inertial in an absolute sense. Any definition is operational—that is, based on experience. We operate on the earth, and Newton's laws generally apply here. Other reference frames moving at constant velocities relative to the earth are considered inertial frames, too; Newton's laws should apply there also. These laws and Einstein's special relativity corrections, however, do not apply in accelerated reference frames and therefore are not universal.

In the special relativity theory, Einstein started from ether experimental dilemmas and the constant speed of light. In the general theory, Einstein had no direct experimental base—only idealized thought experiments of falling elevators and rotating disks. Both are accelerating frames of reference; an elevator accelerates in speed and a disk accelerates in direction. The Lorentz equations do not apply between observers there and on the earth. What principles exist to equate all reference frames? Einstein peopled elevators and disks, and he introduced light rays for clues.

ACCELERATING ELEVATORS AND CURVING LIGHT RAYS

The world appears different in accelerating systems than in inertial systems. If strange forces appear on rotating disks, weightlessness occurs in falling elevators. Most of us have experienced "sinking" sensations in elevators suddenly starting (accelerating) downward. The floor recedes from under us, so that we do not get the full back force on our feet. This is the beginning of weightlessness. The greater the acceleration of the elevator, the less back force we get and the less weight shown on a platform scale under us. If the cable snaps, the acceleration of the elevator and of ourselves is free fall, and there is no back force from the floor. We are weightless. By blowing air downward from the mouth,

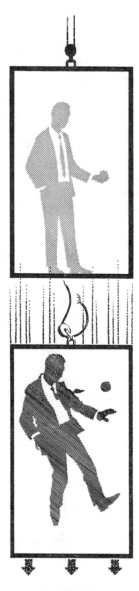

Figure 17.4A

Figure 17.4B

A freely falling elevator is a gravityless world. After the cable snaps, there is no back force from the elevator floor and the man feels weightless. The ball apparently floats with the man.

The floor of an elevator accelerating upward presses strongly against the passenger's feet, the passenger feels particularly heavy, and the ball appears to fall faster than usual. A strong gravitational force seems to operate from the floor.

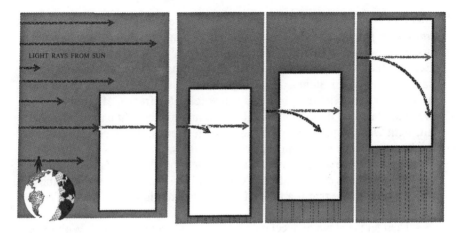

Figure 17.5A

The path of light rays across a stationary elevator is a straight line.

Figure 17.5B

The path of a light ray across an elevator accelerating upward is a parabolic curve. The path is like that of a horizontally fired bullet pulled toward the floor by a gravitational force.

we could even rise to the ceiling of the elevator. A person born and raised in such a freely falling elevator world would not observe any gravitational effects. A ball released from his hand would not drop to the floor; it would float with him (Fig. 17.4A). This is the gravityless world he knows. To an observer on earth, the elevator occupant and objects freely fall together because of the gravitational pull of the earth. What is gravity from the earth as a reference frame is inertia in the elevator. Einstein said that both points of view are correct. What is gravity force and accelerated motion from the earth is not force and motion within the elevator. Force is relative to frames of reference. Gravity forces acting at a distance was Newton's assumption in a great advance of ideas. But since Newton, the mechanical view had run into difficulties.

Suppose that instead of freely falling to earth, the elevator accelerates upward (Fig. 17.4B). The floor presses strongly against the passenger's feet; he feels heavy. He says that there is a strong gravitational attraction between him and the floor of his elevator world. And any objects released from his hand fall to the floor quickly. To an observer on earth, the effects in the elevator are due not to gravity but to accelerated motion. Highly significant to Einstein was that within the accelerating elevator world, there is no way to tell the difference between accelerated motion and gravity effects. Whether up or down in an elevator, acceleration effects are felt as weight effects. Einstein therefore enumerated his famous *principle of equivalence: At any point in space, the effects of gravitation and accelerated motion are equivalent and cannot be distinguished from each other.* This principle also applies to the earth, since it is not an absolute or fixed reference frame, just a convenient frame for us. Because the earth is an inertial frame, laws from it are limited and not necessarily universal. Einstein's deep insight was to use this equivalence principle as a bridge between inertial frames and accelerated frames of reference. This was facilitated by introducing light beams into his thought experiment. After all, light was unique in inertial frames of reference; its speed was absolute. Light may also be unique in accelerating frames.

Now imagine a light ray passing through the glass walls of a stationary elevator above the earth's surface. An observer on earth and one within the elevator would agree that the path of the ray is straight (Fig. 17.5A). The two frames are inertial.

Now let the elevator accelerate upward at 32 feet per second. With respect to the earth, the path of the ray is still a straight line. With respect to the observer in the elevator, however, the path of the light ray is a parabolic curve (Fig. 17.5B), like that of a cannonball fired horizontally on earth. A meteor streaking across the elevator would also form a parabolic curve. To the elevator observer, the floor of the elevator seems to have a gravitational force on the light ray and meteor identical to the gravity of the earth on a stone thrown horizontally. The observer on earth claims that the parabolic curves apparent in the elevator are due to the accelerated motion of the elevator. But there is no way within the elevator world to tell the difference between accelerated motion effects of the elevator. But there is no way within the elevator world to tell the difference between accelerated motion effects of the elevator and any apparent gravitational effects. By assuming gravity, the elevator scientist is assuming what Newton assumed on earth. And there is no preferred or absolute frame. Again, we have an illustration of Einstein's equivalence principle. But this time something significantly new has appeared: *Light in its parabolic curve within the elevator acts exactly like the meteor—like a material substance—regardless of whether the explanation comes from the accelerated motion of the elevator or from an assumed gravitational field within it.*

If this statement is true within the elevator, it is also true elsewhere by the equivalence principle. Light rays, therefore, should gravitationally respond to large masses and be deflected from Euclidean straight-line paths by them. In a 1911 paper "On Influence of Gravitation on the Propagation of Light," Einstein concluded, in effect, that if energy and mass are basically the same, light and other radiation must have mass ($E = mc^2$). If light has mass, then gravity should deflect light. Light, therefore, should not travel in straight lines but in curves unless far from matter in space. If so, the flat Euclidean geometry and its regular straight-line forms do not strictly apply to the universe. Euclidean geometry assumes straight-lines paths for light and no curvature in space.

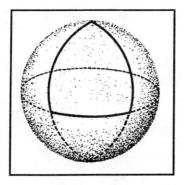

Figure 17.6

A triangle on the surface of an orange has curved sides.

Where light travels in curved lines, a geometry with triangles and other figures whose sides are curves would be more fitting than Euclidean geometry. A triangle on the surface of an orange has curved sides (Fig. 17.6).

So far we have been considering effects in an accelerating elevator. Are there also gravitational effects in circular motion? Could gravity exist for large rotating disks in space far from any other object? Suppose we consider a huge hollow disk rotating at high speed far off in empty space. The disk inhabitants have always been within, and their scientists have formed concepts and theories based upon what is observed from their framework. If the spaceship disk moved at constant velocity relative to the stars and rotated slowly like the spaceship earth, their laws and theories would be similar in character to those on earth. But suppose that the spaceship rotates quickly. An inhabitant takes for granted the inertial or centrifugal forces—from the earth's point of view—pressing him against the side of the disk. (The disk's side is actually his floor.) To him, the force would be characteristic of his world. He might even call the force gravity, since its effects would be the same. His Newton may even have established a mathematical law of gravitational attraction. What are called centrifugal effects from Newtonian laws on our inertial frame are called gravity or attraction effects from concepts and laws on his rotating frame. And so Einstein's idea of equivalence also applies to circular motion.

This equivalence principle is now taken for granted in space travel. Einstein's elevators are today's spaceships. Consider astronauts in a rocket far out in space. As long as the rocket travels at constant velocity, the astronauts are weightless. No large body is near enough for the gravitational effect called weight. If the rocket suddenly accelerates forward, the astronauts are pressed

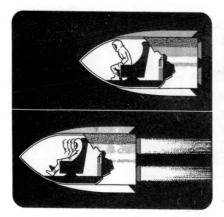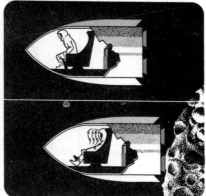

Figure 17.7

Acceleration of the rocket forces the astronaut back in his seat just as if a huge asteroid had suddenly appeared behind him.

back in their seats by inertia just as if by gravity (Fig. 17.7). The effect on the astronauts would be the same as if a huge asteroid suddenly loomed behind them. Einstein spoke of gravitational fields and equivalent inertial fields. To him fields were real; his main professional objective was to unify all fields—gravitational, inertial, and electromagnetic.

For future space travel, Einstein's principle of equivalence is taken for granted. Spaceships will spin considerably while moving toward their destinations. Rotational effects keeping passengers along the sides of enclosures will be equivalent to gravitational effects. Weightlessness would be overcome. The sides of the enclosures will be designed as the floors of rooms. Although originating from thought experiments, Einstein's equivalence principle is here to stay.

ROTATING DISKS, LAGGING CLOCKS, AND CURVING RULERS

Disks rotating at high speed show distortion of space and time within. The farther from the center of the disk, the faster a point on it moves. The direction of motion at any point is perpendicular to the radius. By the Lorentz transformation principle, objects shrink in the direction of motion: the higher the velocity, the more the shrinkage. Therefore, a very small shrinkage existing near the center of the disk at a tangent increases with distance from the center. On the other hand, no shrinkage exists in the direction of a radius, since

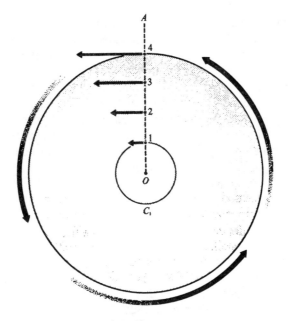

Figure 17.8

The further a point is from the center of a rotating disk, the faster the point moves and the greater is the disk's shrinkage.

that is perpendicular to the direction of motion. An observer on the rotating disk should find that the circumference C_1 of a small circle around the disk center almost equals $2\pi r_1$* ($C = 2\pi r_1$); a large shrunken circumference C_4 does *not* equal $2\pi r_4 (C \neq 2\pi r_4)$. The shape of the disk, the space on it, is distorted. A basic proposition in Euclidean geometry is that $C = 2\pi r$. Einstein was convinced that Euclidean geometry does not apply to accelerating frames of reference (Fig. 17.8).

Now place synchronized clocks on a radius of the rotating disk. By special relativity transformation, the farther the clocks are from the center, the more they slow down. Increased velocity slows clocks. Time, too, is distorted on the disk (Fig. 17.9 and 10).

In summary, by Einstein's principle of equivalence, gravitational mass effects are equivalent to accelerated motion effects. The disk thought experiment represents accelerated motion. Therefore, if rotating disks show distorted space and time, then large masses, such as the sun and stars, should

---

*Where r is a radius.

show distortions of space and time around them. Euclidean geometry, therefore, would not apply close to large masses. And light, like matter, would also take a curved path on a rotating disk and near the sun and stars. All this from thought experiments about elevators and disks!

But if ordinary "flat" geometry does not apply to irregularities of space, then what kind of geometry does? Such great mathematicians as the Hungarian John Bolyai (1802–1860), the Russian Nicholas Lobachevsky (1793–1856), and the German G. F. B. Riemann (1826–1866) had developed non-Euclidean systems of geometry, but no scientist until Einstein had seriously applied non-Euclidean geometry to the physical universe.

GEOMETRY, AN ORANGE, AND A SADDLE

In Euclidean geometry, the sum of the angles of a triangle equals 180°. Like all statements, this one has assumptions and limitations. A basic assumption here is that the triangle is on a flat surface. The sum of the angles of a triangle drawn on an orange equals *more* than 180° (Riemann, Table 17.1). Or, on certain areas of a saddle, *less* than 180° (Lobachevsky, Table 17.1). The smaller the orange, the larger the sum of the degrees of the triangle. The larger the orange, the closer the angles approach a 180° sum. A small triangle on an orange the size of the earth would have the sum of its angles very slightly more than 180°. And so space even on the earth's surface is not what it seems.

On a spherical surface like the earth, an arc of a great circle, or longitudinal line, is the shortest distance between two points, not a Euclidean straight line. Great circles or "polar routes" for planes are easier in the nature of things than digging tunnels through the earth between various points on the surface. *Geodesic* is the term for a line that is the shortest distance between two points on a surface or in space, regardless of the shape of the line. At the equator, great circles, or lines of longitude, are parallel and, according to Euclid, should not meet. By the nature of the earth's spherical surface, however, these lines do meet at the poles, the sum of the angles of a triangle drawn anywhere on the earth's surface is more than 180°, and $C \neq \pi D$ (or $C/D < \pi$), as emphasized in Riemann's geometry.

In his geometry, Lobachevsky used a pseudosphere—for example, a saddle (Table 17.1)—as a model. He proved that in cases of such models where *two* lines can be drawn through a point parallel to a given line, the sum of the angles of a triangle is less than 180°, and $C/D > \pi$. Euclidean "flat" geometry, with its straight lines and 180° triangles, applies to space that is not curved; Riemann's "spherical" geometry, to positively curved reaches of space; and Lobachevsky's geometry, to negatively curved space.

Table 17.1
Flat versus Curved Surfaces

	EUCLID (EUCLIDEAN GEOMETRY)	RIEMANN (ELLIPTICAL GEOMETRY)	LOBACHEVSKY (HYPERBOLIC GEOMETRY)
Fifth Postulate	One and only one line can be drawn through a point parallel to a given line.*	No line can be drawn through a point parallel to a given line.	Two lines can be drawn through a point parallel to a given line.
Surface	Plane	Sphere	Pseudosphere
Curvature	Zero	Positive	Negative
Triangles			
	The sum of the angles of a triangle equals 180°.	The sum of the angles of a triangle is greater than 180°.	The sum of the Angles of a triangle is less than 180°.
Ratio of Circumference to Diameter	Pi (π)	Less than pi ($<\pi$)	Greater than pi ($>\pi$)

*This is an equivalent statement to Euclid's fifth postulate.

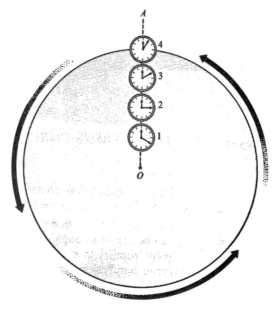

Figure 17.9

Increased velocity slows clocks. The farther a clock is from the center of a rotating disk, the more the clock slows down.

But what could be meant by curved space? And how could such space be detected? Light enters on the second question. If space is curved, light would follow the curve and reveal it just as in an accelerated elevator. If, for example, space is curved around the sun because of its large mass, then light from a distant star would curve around the sun. Einstein's principle of equivalence predicts that gravity affects the path of light. Deflection of light should be observable during a solar eclipse and should involve triangles with curved sides, as on the earth's surface (Riemann's geometry). Euclidean geometry, of course, would be inappropriate. But the first question, the *nature* of curved space, is more difficult.

Take points A and B near the sun. A ray from star S would be deflected form a straight-line path BC' to curve around the sun to BC (Fig. 17.10). A ray from another star T would be deflected near the sun from BA' to BA. A third ray from star U would take the curved path AC. The triangle formed would clearly be of the Riemann type: the sum of the angles would be more than 180°.

Space is appreciably distorted on very fast disks and near large masses. Nor is time uniform on disks rotating speedily. By the underlying principle of

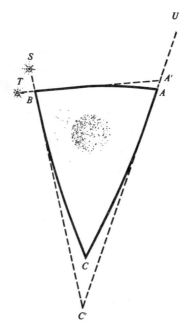

Figure 17.10

Curvature of the paths of starlight around the sun. The triangles formed are like those on the surface of a huge orange.

equivalence, clocks or any rhythmic process should be slowed down in small amounts in gravitational fields near large masses.

Euclidean, or flat, geometry does not apply to rotating disks. The shortest distances on them are not Euclidean straight lines but curves. What kind of curves? Take points *A, B,* and *C* on a rotating disk (Fig. 17.11). When the disk is stationary, an observer on it would find that the shortest distance would be the dashed lines *AB, BC,* and *CA*. The smallest number of rulers he would need end to end would be along Euclidean straight lines. Once the disk rotates, an outside observer should see the inside observer curving in from *A* to *B, B* to *C,* and *C* to *A* as he seeks to triangulate with the smallest number of rulers. The explanation is relativistic contraction of length of the rulers: Moving a ruler closer to the center gives it a smaller linear velocity and therefore less shrinkage. A ruler would cover more ground by curving inward. Up to a point, the added distance from the point of view of an outside observer is more than balanced by reduced shrinkage. A curve becomes a shorter distance. And the

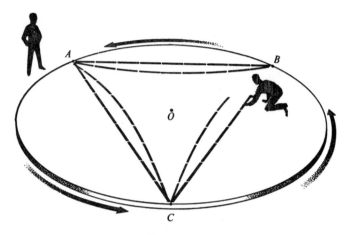

Figure 17.11

What appears as a straight line of rulers to an observer on a quickly rotating disk appears as a curved line to an outside observer.

faster the disk rotates, the more the bending inward for the geodesic. The curve in this case would be Lobachevskian; the sum of the angles of the triangle would be less than 180°. And since light also propagates along the shortest path, light would show the nature of the curve or the space.

To further illustrate that Euclidean measurements do not apply in non-Euclidean space, consider the street plans in Chicago and San Francisco. The Chicago area is flat, its street plan highly rectangular and regular. Madison Street is a straight-line axis separating the north side from the south side. State Street is a straight line separating east from west. Almost all city streets are parallel to these two, and all numbering starts from the two axes. The city street plan has the uniformity of a Cartesian coordinate system (Fig. 17.12A) based on "flat" Euclidean geometry. Some streets in San Francisco follow the contours of its high hills, at least in part. To force the flat Cartesian plan, excellent for Chicago, upon San Francisco would be folly. It would mean streets tunneled into hills. Curved surfaces are often more adequately handled and numbered by a Gaussian coordinate system of wavy distended lines adjusted to fit specific situations. Figure 17.12B illustrates a Gaussian system in that the x and y coordinate lines (the white lines) are not parallel to each other.

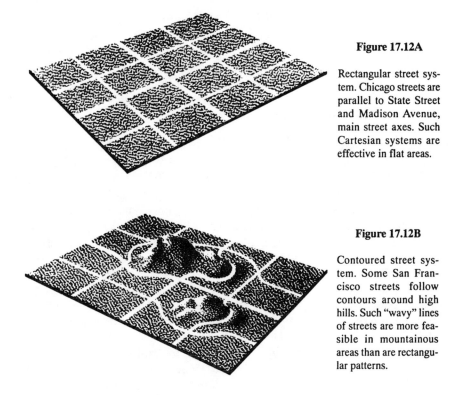

Figure 17.12A

Rectangular street system. Chicago streets are parallel to State Street and Madison Avenue, main street axes. Such Cartesian systems are effective in flat areas.

Figure 17.12B

Contoured street system. Some San Francisco streets follow contours around high hills. Such "wavy" lines of streets are more feasible in mountainous areas than are rectangular patterns.

WORLDS OF DIFFERENT DIMENSIONS

The mildly curved space in the universe is "shaped by masses and their velocities," said Einstein. "The geometrical properties of space are not independent, but are determined by matter." Basically, space is a sequence of objects, and time a sequence of events. Without objects or events, space or time does not exist. There is simply nothing, a zero with the rim knocked off. But objects endure in time and therefore become events. Just as electric fields and magnetic fields become electromagnetic fields, separate space and time are inseparably linked in Minkowski's concept of space-time. Therefore, space distortions and time distortions around large masses become combined space-time distortions, a four-dimensional concept. And Einstein preferred to consider gravity as a continuous field in four-dimensional space-time rather than as a Newtonian force acting at a distance in three dimensions of space separate from time. Difficulties in understanding this, said Einstein, may arise from the fact that we are three-dimensional creatures in a four-dimensional world and must compensate mathematically for our physical limitations. Two lines of sight from two eyes give us depth, the third dimension of vision. One eye would give us only two dimensions. Physiological, mechanical, and astronomical rhythms give us a separate awareness of time. Our physiological mechanisms do not combine space and time. Einstein suggests the following analogies of one-dimensional creatures in a two-dimensional world and two-dimensional creatures in a three-dimensional world.

A one-dimensional creature (Fig. 17.13A) in a one-dimensional world is confined to a straight line. Without sight, it is limited to that line with a "feeler," much like a streetcar to an electric line by a trolley. The one-dimensional creature can only move forward or backward; sideways or up and down do not exist in its vocabulary. If its straight line has terminals, the one-dimensional creature lives in a *finite* and *bounded* world—finite because the line has a fixed length and bounded because of the stopping points. Without terminals, the one-dimensional creature would have an infinite and unbounded world, even though of a single dimension.

But suppose our one-dimensional creature were placed in a two-dimensional world, a circle (Fig. 17.13B). The circle is so large that, like an earthling going around the equator, it may believe it is always on a straight line. With the best of intelligence, it would be difficult for the one-dimensional creature to understand how it could always move forward in one direction and eventually find itself back where it started. Visualization of a second dimension would not happen by everyday experience alone, but only by mathematical imagination. If the one-dimensional creature had previously experienced a circle as a mathematical game of two dimensions, it could transcend its one-dimensional limitations and explain how it could move in only direction and yet return to the starting point. It would understand how its world could be *finite* (the circumference of a circle) and still be *unbounded* (no end points).

A two-dimensional creature in a two-dimensional world would be much like a creature on a flat movie screen. Thickness would not exist, just length and width (Fig. 17.14A). The world would really be flat, and the creature one-eyed. Circles could be seen and understood, but spheres would be seen only as circles. Euclidean plane geometry could exist and apply to the world, but Euclidean solid (three-dimensional) geometry would only be only a game of logic for imaginative mathematicians. As a huge screen, the world would be finite and bounded. If our one-eyed creature was rooted to the earth, railroad tracks would appear as coming to a point, and trains as monsters that puffed up or shrank in size rather than approached or departed. Even Euclidean plane geometry would not directly apply to its observable world if the two-dimensional creature observed parallel lines as coming to a point.

Now place a movable two-dimensional creature on curved screen, in fact

A – FINITE & BOUNDED

Figure 17.13

(A) One-dimensional creature in a one-dimensional world.
(B) One-dimensional creature in a two-dimensional world.

B – FINITE & UNBOUNDED

on a gigantic spherical screen, and remember that thickness is not observable to the creature (Fig. 17.14B). Only a mathematical two-dimensional creature could visualize how he could start off in any direction on his spherical screen in a straight line and land at his starting place, or how the screen world could be *finite* (a limited measurable surface area) but *unbounded* (without boundaries on that area). Once again, mathematics would transcend the physical limitations of an observing creature.

Einstein's point, of course, is that if we are three-dimensional creatures in an assumed four-dimensional space-time world, we are physically handicapped. We visualize in three dimensions, not four. A mathematical model involving non-Euclidean (curved) geometry and flexible, wavy Gaussian coordinate systems could overcome physical limitations.

Gravity in a Four-Dimensional World

Newton's gravitational force between two masses depended upon the distance between them ($F = G\,Mm/d^2$). That is, the gravitational force equals the gravitational constant G times the product of the two masses $M + m$ divided by the distance squared (d^2) between the two masses. Masses and the distance between them had the same values for all observers. Matter had mass; energy did not. The gravitation idea had successfully explained the heavens and worked marvels on earth—with one serious exception. The theory could not fully explain why the orbit of Mercury slowly rotates around the sun about one-sixth of 1° a century in a rosette pattern.

Could the closeness of the sun to Mercury be a disturbing factor? The sun

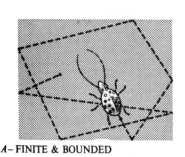 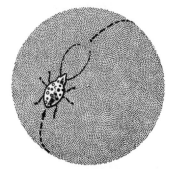

A– FINITE & BOUNDED   *B*– FINITE & UNBOUNDED

Figure 17.14

(A) Two-dimensional creature in a two-dimensional world.
(B) Two-dimensional creature in a three-dimensional world.

is massive, and huge masses have effects equivalent to those of accelerated frames of reference. Light rays curved in Einstein's accelerating elevator (Fig. 17.4B), whether due to gravity as the passenger thought, or to the elevator's motion, as the earthly observer thought. The earth is an inertial frame of reference; the sun is not. The sun's huge mass would give its immediate surroundings effects equivalent to those on a quickly rotating disk or in an accelerating elevator. Newton's laws would not apply, and understanding Mercury's rotating orbit requires a concept of gravity appropriate near massive objects.

What could such a gravitational concept be? A basic postulate of optics is that light always takes the shortest path. In inertial frames and near small masses, the shortest path is a Euclidean straight line or very nearly so. But near large masses, light should curve just as on a rotary disk or in an accelerating elevator. These light rays curves are the shortest paths, or geodesics, just as great circles arcs on earth are the shortest distances between points. That is, light rays or meteors, in taking these curved paths, are taking the easiest routes and are therefore in natural motion; on forces are acting on them (Fig. 17.15). Einstein reverses the picture: Instead of saying that gravity is a force acting equally on material objects and light to give them the same curved path, he argues that no forces act in either case.

But if no forces act, then why the *curved* path of objects and light near large masses—and what, for that matter, is gravity? The answer is: *field*. Gravity is a field, not a force acting at a distance. That answer returns us to space-time and curved (Gaussian) coordinate systems. Einstein explained the curved path of a ray as an effect of space-time distortions around masses that is noticeable when masses are large. The sun is accompanied by distortions in space and time form-

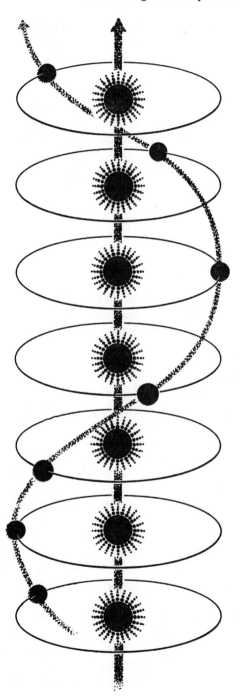

Figure 17.15

The world line, or natural path, of the earth in space-time.

ing space-time fields, much like moving magnets surrounded by electromagnetic fields. Such space-time distortions or gravitational fields structure our four-dimensional space-time universe. The four-dimensional gravitational field exists whether objects and light wander in or not.

When a meteor or light ray does wander in, it is guided by the four-dimensional gravitational lines, or "grooves," already there. The particular line becomes the shortest path of the objects or ray in space-time. Far from the sun or any mass, the "guiding lines" become Euclidean straight ones. All these "guiding lines" are considered as forming a continuous pattern or model called a "space-time continuum," whether "straight" and flat in empty space or "curved" where matter is concentrated. If matter is properly concentrated in the universe, light can curve around enough to return to a starting place and provide a glimpse into primeval history.

FREE BODIES AND STRUCTURED PATHS

The entire discussion in this chapter builds up to *Einstein's one general law of motion: All free bodies pursue geodesics, or straight paths, through space-time regardless of whether they are near matter or remote from it.* This law is a combined law of motion and gravitation that relates all frames of reference, accelerated and inertial. When the "free body" is remote from other matter, the geodesic follows a Euclidean straight line in four-dimensional space-time and also in three-dimensional space if observed from an inertial frame of reference. If the body is near matter, the geodesic is curved in space-time and, therefore, also in space. The earth moving around the sun is therefore a "free body" in the same sense that a meteor is in far-off space. Each follows the space-time structure in its vicinity. Space-time is curved near the sun. The natural path, or world line, of the earth in space-time is therefore curved (see Fig. 17.15). Far from matter, the world line of the meteor is straight. Thus, in general relativity, a force is not necessary to explain the orbit of a planet or moon. A planet or moon follows a structured a path as the natural, easiest path, whether near or far from matter. Gravity is the four-dimensional field in space-time structuring the paths of objects and radiation. These paths, world lines, or geodesics are curved near matter, less curved away, and straight (Euclidean) far off in empty space in accordance with the distortions of the gravitational field by large masses. But curved or straight, the paths are continuous. *Briefly, gravity is a continuous field in space-time, influencing and being influenced by matter.* The special theory of relativity is a special case of the general theory in which space-time paths, undistorted by large masses, are straight in a Euclidean sense and gravitational influence is negligible.

The huge space-time continuum is an absolute connecting everything in the universe. Individual measurements of separate space and time are relative, as emphasized in Einstein's special theory; each observer has his own local values. But an Einstein interval, or distance between point-events on the world line of an object, is the same for all observers everywhere. An absolute four-dimensional space-time continuum was the scientist Einstein's answer to the artist Goethe's conviction, in an essay he wrote, that "in nature we will never see anything isolated, but everything in connection with something else which is before it, beside it, under it, and over it." There was much of the artist in Einstein's work, just as there was the scientist in much of Goethe's.

ONCE AGAIN, SEEING IS BELIEVING

Does Einstein's space-time model really fit the universe? What is the evidence?

If gravity affects the path of light, the sun's mass should deflect starlight that grazes the sun. To test this prediction of Einstein, A. S. Eddington (1882–1944), one the world's great astronomers, led a British astronomical expedition to the West African island of Principe to observe the solar eclipse of May 29, 1919. Another British expedition, under A. C. Crommelin, went to northern Brazil for the same purpose. Ordinarily, the sun is too bright for starlight to be seen. During a solar eclipse, however, when the moon blots out the sun, starlight grazing the sun can be photographed. Since the relative position of a given star to other stars is known, the bending of its rays is easily determined. Einstein's four-dimensional gravitational field equation gave 1.74" of arc as the deflection of a light beam grazing the sun. Newton's law of gravitation ($F = GMm/d^2$) gave a deflection value of 0.87" of arc. The relativistic mass m of light energy E ($E = mc^2$) was used for m in Newton's equation.

Einstein stated in advance that the eclipse test was crucial for his general relativity idea of space distortion by masses even if not for his special relativity concept of $E = mc^2$. If there was no deflection of starlight at all, the general theory of relativity would be invalidated, and the $E = mc^2$ matter-energy equivalence principle would be seriously shaken. If a deflection near 0.87" arc was found, the general relativity theory of gravitation would be out, Newton's concept of gravitational force at a distance strengthened, and $E = mc^2$ upheld. If a deflection near 1.74" of arc was found, Einstein's concept of gravity as a field in space-time deserved further exploration, for it would show far more accuracy and scope than Newton's concept. Scientific excitement ran high in May 1919 and even higher when solar expedition photographs gave average deflections close to Einstein's predicted value. Other solar eclipse data since 1919 have further confirmed Einstein's space-time concepts of gravitation (Fig. 17.16).

We have already noted that the slight rotation of Mercury's orbit could not be fully explained by Newtonian gravitation. The gravitational effect of known

Figure 17.16

Evidence of general relativity: the bending of starlight around the sun.

planets accounted for only 531" of the 574" (about ⅙°) of arc of this rotation. A new planet, Vulcan, was hypothesized as being near Mercury. But Vulcan was never found. The excess rotation of Mercury's orbit remained unexplained until Einstein's relativity. Mercury increases its speed when closer to the sun than when farther away. Einstein's special relativity predicts a variation of Mercury's mass with its speed. This variation of mass accounts for seven of the forty-three unexplained seconds of arc per century. General relativity mathematics accurately accounts for the remaining 36" of arc. This successful explanation of the rotation of Mercury's orbit is perhaps the best evidence so far for Einstein's theory of gravity (Fig. 17.17).

A clock slows down with its increased velocity whether in an inertial reference frame, in an accelerating frame of a rotating disk, or in an accelerating spaceship. By Einstein's principle of equivalence between accelerated motion and gravitational effects, time should also slow down appreciably on or near gravitational masses. If time "stretches," then space-time should, too. Einstein figured that a second on the sun should be 1.000002 earth seconds because of the sun's greater mass. Jupiter's mass is larger than the earth's but smaller than the sun's. A second on Jupiter should also between that of the earth and that of the sun in length. Vibrating atoms act as clocks. Vibrations of hydrogen atoms, for example, on the sun or stars should be slower than hydrogen atom vibrations on earth. This can be tested by color analysis of starlight dispersed in prisms. Glowing hydrogen gas has its characteristic colors. If vibration frequencies of hydrogen atoms decrease, the colors due to hydrogen will shift toward the red, or lower frequency, end of the spectrum. This reduction in frequency, due to the presence of large masses, is known as the *relativistic* or *Einstein red shift* to separate it from red shifts due to other causes. The Einstein shift was first found on white dwarf stars much more dense than the sun. Sirius B, twin star of the Dog star, Sirius, with a diameter about one-thirtieth that of the sun, has a density over 25,000 times greater. A handful of material from the interior of that star would weigh tons. The astronomer Herbert Adams, in

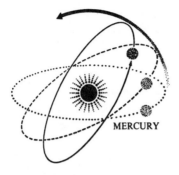

Figure 17.17

More evidence of general relativity: rotation of Mercury's orbit around the sun. The orbit of the planet Mercury very slowly rotates in a rosette pattern around the sun at a speed of less than 1' of arc per century, close to Einstein's prediction.

1925, was the first to find a relativistic shift on Sirius B that matched the predicted amount.

Among other evidence for the slowing down of time processes by gravitational fields is an experiment conducted at Harvard in 1960. Two extremely precise "atomic clocks" were placed at the base and top of a 70-foot tower. As expected from general relativity predictions, the time on the ground was very slightly slower than the time 70 feet above the earth's surface; the earth's mass definitely affects time!

THE SIGNIFICANCE OF RELATIVITY

Perhaps more time is needed to realize the full impact of relativity. For better or worse, our atom-smashing age substantiates the special relativity theory. Accelerated particles increase in mass; matter is converted into energy; and, with it, atomic bombs loom heavily over us all. As for the general theory of relativity, evidence is still sparse but favorable. More time is needed for solid substantiation.

Meanwhile, relativity compels us to ask: If such basic and solid assumptions as Newton's absolute time and space could go toppling, what assumptions do we hold today that the future will also cast aside? Such a possibility justifies our questioning all assumptions, definitions, concepts, and systems of scientific as well as other ideas. Also justified is that type of thinking which, while encouraging free play of the imagination in "calling the world to order," still insists upon testing imaginative ideas.

Relativity is now a conceptual model that has absorbed Ptolemaic and Copernican-Newtonian models into itself as partial pictures. The earth is an organizing center from which thinking men worked their way from immediate, private, relativistic pictures of the universe to a common space-time model. To quote again from Minkowski: "Space by itself and time by itself are doomed to fade away into mere shadows and only a kind of union of the two retains self-independence." But, of course, shadows are part of reality, too. The Ptolemaic and Copernican-Newtonian models are three-dimensional partial pictures of Einstein's larger four-dimensional model. Men can work outward only from where they are. From the earth as a frame of reference, geocentric (Ptolemaic) models were in order for sea and air navigation. But for space travel, the sun is a basic frame of reference, and sun-centered (Copernican) models are preferable. But difficulties arise is the relativism of different frames of reference. From the earth, the orbit of the moon is nearly a circle; from the sun, it is not. Hence, an orbit seen as circular from one frame is not so from another. In an absolute sense, there is no "correct" frame. Where everything moves, absolute velocities are unknown and there is no preferred frame of reference for universe laws. Observers on the earth, sun, or anywhere, however, can have common world lines of the moon in space-time. The space-time model is a larger reference relating all observers.

In this way, through grand, unifying ideas, scientific man makes his way outward in a search for order in a complex, changing world. His big ideas run into contradictions, face more promising ideas, and are eclipsed. But each big idea has its own expansive effect. The broadening effect of Einstein's model in specific concepts of space, time, matter, energy, gravity, or geodesics has already been described; so has Einstein's frames of reference approach to the universe. It is the best so far. In turn, it may become a facet in a still larger framework; but meanwhile, like previous models, Einstein's model helps man creatively project himself outward—and gain further insights into the process of interaction with his surroundings.

SUGGESTIONS FOR FURTHER READING

D'Abro, A. *Evolution of Scientific Thought—From Newton to Einstein.* New York: Dover, 1950.
Davies, Paul. *Superforce, Search for a Grand Unified Theory of Nature.* New York: Touchtone, 1985.
Eddington, Arthur. *Space, Time and Gravitation.* New York: Harper, 1959.
Hawking, Stephen W. *A Brief History of Time.* New York: Bantam Books, 1988.
Kaufmann, William J., III. *Black Holes and Warped Space-Time.* New York: Bantam Books, 1979.
Landau, L. D., and G. S. Rumer. *What Is Relativity?* New York: Dover, 1977.
Rucker, Rudolf. *Geometry, Relativity and the Fourth Dimension.* New York: Basic Books, 1977.
Russell, Bertrand. *The ABC of Relativity.* New York: Mentor, 1959.
Schilpp, Paul Arthur. *Albert Einstein, Philosopher-Scientist.* New York: Tudor, 1951.
Will, Clifford M. *Was Einstein Right? General Relativity Tested.* New York: Basra Books, 1986.

18

Measurement as Intervention:
Heisenberg's Uncertainty Principle

> An intelligence knowing, at a given instant of time, all forces acting in nature, as well as the momentary positions of all things of which the universe consists would be able to comprehend the motions of the largest bodies of the world and those of the smallest atoms in a simple formula, provided it were sufficiently powerful to subject all data to analysis; to it, nothing would be uncertain, both future and past would be present before its eyes.
>
> Pierre Laplace, *Celestial Mechanics,* 1825

Twentieth-century developments in photon and quantum theories have severely shaken the deterministic certainty of Laplace's words quoted above.

At the end of the nineteenth century, light seemed entrenched as electromagnetic radiation emitted in continuous waves by vibrating electric charges in luminous bodies. Newton's color spectrum was now part of a far more extensive electromagnetic spectrum (Fig. 18.1). But sooner or later, situations arise that the best of conceptual models cannot explain. Just when the electromagnetic wave theory seemed most firm, a *photoelectric effect* was observed that contradicted it. Light falling on a metal plate ejected electrons (negative charges) from the plate (Fig. 18.2) to positively charge the plate. How is it possible for light waves to have an impact on electrons? Water waves impact on objects, but water is a material thing. Perhaps light is not "pure" energy after all!

The difficulty really began, however, when the intensity of the light on the photoelectric plate was increased. Instead of electrons leaving the plate with *more* speed, there were simply *more* electrons leaving at their old speed. The wave theory of light should mean higher waves with nastier "wallops" on plate electrons: the electrons should move more quickly. But if light consists of bulletlike particles, more intense light would mean more "bullets" at the same speed. A given electron hit by a light particle would not be affected by the addi-

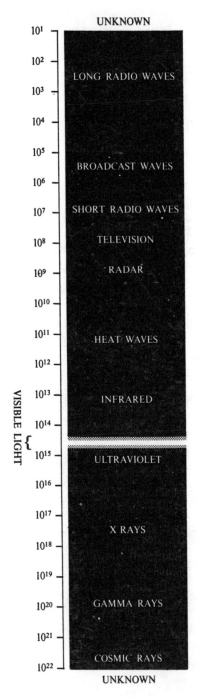

Figure 18.1

The electromagnetic spectrum. Visible light (3,800 angstroms to 7,500 angstroms) is a narrow, 1-octave range of color in a 60-octave band of electromagnetic waves stretching from long radio waves to cosmic rays.

tional "bullets." More "bullets" would simply mean that more electrons would be hit and emitted. From the results, light seemed more like bullets. So maybe light is not a wave phenomenon after all!

In any case, Albert Einstein (1879–1955) assumed in 1905 that light is a stream of energy particles he called *photons*. Einstein's photons are not minute bits of matter like Newton's corpuscles; they are particles of *energy* that act like matter. The energies of Einstein's photons are directly proportional to the color frequencies given in the electromagnetic spectrum. Violet light has a higher frequency than red light. Therefore, photons of violet light are greater in energy than red light photons. The larger photons of violet striking metal should cause electrons to move faster than the smaller red photons do. Experiments showed this to be so. In his photon idea, Einstein was applying to light an energy packet, or *quantum* idea, that the German physicist Max Planck (1858–1947) had successfully applied to heat phenomena. Einstein assumed that the energy E of each particle of energy was represented by $E = hf$ (Equation 18.1), where h is a now famous universal constant of nature called *Planck's constant* (with a value determined at 6.62 X 10^{-34}, joule-seconds*), and where f is the frequency of the particular radiation in the electromagnetic spectrum.

Note that Equation 18.1 combines photon *discreteness* through Planck's constant h with the *continuity* of electromagnetic spectrum frequency f. The contradictory opposites of discreteness and unbroken continuity indicate duality[1] of photon energy in the equation.

In the transfer of a photon's energy to an electron in the photoelectric effect, Einstein emphasized that $hf = \frac{1}{2}mv^2 + w$ (Equation 18.2), where hf equals the energy of the light photon, $\frac{1}{2}mv^2$ is the kinetic energy of the expelled electron, and w is the energy necessary to remove the electron from the metal. This famous photoelectric equation earned for Einstein a Nobel prize. Equation 18.2 reconciles photon discreteness of energy on the left of the equation with classical physics energy concepts following the equal sign.

Today, photoelectric cells (popularly called "electric eyes") open supermarket doors for customers, but the photon theory that explains this operation cannot explain diffraction phenomena. Bulletlike photons, in passing through a single pinhole, should create a spot on a screen directly in line with the hole. But alternate light and dark rings are also found. And photons through two pinholes should show two spots on a screen. Instead, a series of alternate light and dark stripes are found around a single spot midway between the two pinholes. The photon theory is also unable to explain how light bends around obstacles.

---

*A *joule* is a unit of energy in the metric system equal to the work done by a force of one neutron through a distance of one meter. *Joule seconds* are the product of energy used and the seconds of time during which they were used.

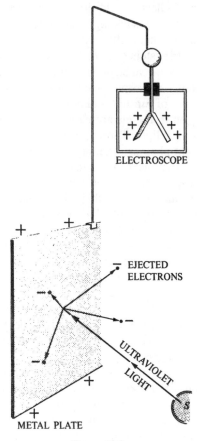

Figure 18.2

Photoelectric effect. Light falling on a metal plate ejects electrons from the plate, to charge the plate positively. The electroscope leaves repel each other, indicating that the plate is charged.

But the opposing wave theory cannot explain the photoelectric effect. The nature of light thus seems relative to the experimental setup. Light is wavelike in a pinhole situation and particlelike in a photoelectric cell. But how can light be both a continuous wave and a discontinuous particle? Perhaps waves and particles are aspects of a bigger picture. Different situations may display light in its different aspects, just as at different angles "the neck of the peacock shows different colors."

The quantum idea began in the year 1900 with Max Planck. Niels Bohr, however, in 1913 sought answers for origins of energy bundles within atomic structure itself. He limited orbiting electrons to a set of allowable circular orbits. Electrons abruptly jumped between designated orbits in line with patterns of spectral lines emitted by hydrogen atoms. The energy that the electron lost in falling from a given level toward an allowable level closer to the atomic nucleus would be equal to the energy of emitted light from the hydrogen atom somewhat analogous to energy transfers of objects falling from set heights due to gravity. Within atoms, energy values are fixed due to electron jumps between set energy levels around the atomic nucleus. Thus, Bohr gave us insight into quantum energy bundles, which was not answerable simply to the continuity of Maxwell's electromagnetic waves.

Discrete bundles of energy for hydrogen were confirmed by experimental evidence a year later, in 1914, by German physicists James Franck and Gustav Hertz. Hydrogen is the simplest atom. Unfortunately, similar predictions for chemical elements beyond hydrogen were not experimentally successful. Quantum theory needed further development such as in Schroedinger's wave mechanics.

But let us return to Max Planck, who was able to introduce his constant h called Planck's constant. This constant could accurately explain heat radiation emitted by an idealized model of a black body that absorbs all incoming radiation and then reradiates it in a dark room in a characteristic pattern first predicted by Planck. The significance was that the radiant energy was never less than Planck's constant h times the frequency f of atomic oscillations. In his famous radiation law, Planck assumed that the radiant heat energy E emitted by a heated surface equalled h times f, or $E = hf$. Since h is an irreducible constant, the energy E radiated was only in set, discrete bundles that Planck called *quanta*. The quanta were emitted when an oscillating particle changed from one energy state, E_1, to another, E_2, or the radiation emitted $hf = E_1 - E_2$.

THE HEISENBERG UNCERTAINTY PRINCIPLE

As the saying goes, truth is often stranger than fiction. This is especially applicable to science. We have seen (chapter 12) that in the process of observing, we interact with what we observe. We know, for example, that color is a sensation in response to light radiation. When the incoming wavelength of light from a flower is about 3,800 angstroms, the flower appears violet. At a wavelength of about 7,500 angstroms, a rose appears red. The sensation of red, an *interaction* between the eye mechanism and the incoming radiation, appears as a characteristic of the rose. Without an eye, there is no color, only light waves, just as without an ear, there is no sound, merely sound waves or vibrations of air or other molecules. Without a world of objects, there is nothing to view, nothing to originate or reflect radiation or other vibrations. Without sensory organs there are no responding sensations to incoming energy vibrations. The sun, the eye, and a world of form and color go together. The process is interactively creative.

But with quantum theory and Heisenberg's uncertainty principle, the relationship between an observer and what is observed can be carried from a process of *interaction* to one of *intervention.*

Photoelectric effects and Einstein's photoelectric equations (Equation 18.2) attested to photon bundles of energy packing a "wallop" upon electrons on metal plates (Fig. 18.1), with photons acting like matter on the electrons. Werner Heisenberg came to a highly significant scientific and philosophic conclusion in his *uncertainty principle* (1927), also known as the *indeterminacy principle*: there can be no observation without light photons or other radiation. Since the "wallop" by a photon or quantum of energy affects the velocity of an electron indeterminately at any point, there would have to be some unpredictability or uncertainty in knowing simultaneously the exact position and exact velocity of the electron. And the electron cannot be seen without the "walloping" photon. That is, there is no way in nature of measuring the exact position and the exact velocity of an object at the same time, especially since light and other radiations have dual characteristics of both waves and particles. Heisenberg limited this uncertainty in the following equation:

$\Delta x \Delta p = h/2\pi$, where Δx is the uncertainty of position,

Δp is the uncertainty of momentum,

h is Planck's constant, and

$h/2\pi = 10^{-34}$ joule-seconds, a very minute value.

The $h/2\pi$ is a constant value. Therefore, the smaller the error Δx in the position of the electron, the larger the error Δp in the momentum. Or, the more exact the determined position of the electron, the greater the error in simultaneously determining the velocity of the electron. And vice versa. The *product* of the two uncertainties of position (Δx) and momentum (Δp) is always the same, $h/2\pi$. The combined error cannot be greater than $h/2\pi$, so minute in value that it becomes significant only in the atomic and subatomic world, with its very small masses that may be impacted by photons. The impact of a photon on an automobile in the everyday world is, of course, negligible.

The significance of even a limited uncertainty in this subatomic realm opened the doors to statistical or probability approaches in physics. More was needed than the highly deterministic approaches of classical physics. At least in twentieth-century subatomic physics, the old complacency could no longer remain dominant.

Also significant is the observation that in investigating nature, science often *intervenes* as well as interacts. Is it possible to intervene and interact and still be *totally* detached and objective? Furthermore, is not science, like other cultural aspects, dealing with its own partial images and associated practices along with other limitations of nature and knowledge?

Intervention involves interaction between apparatus and what is observed. With diffraction gratings, light has wave characteristics. With photoelectric cells, light shows contradictory particle characteristics. More than one hypothesis or theory can at times explain phenomena where different techniques are used—until more knowledge and/or a more all-embracing theory is found. Contradictory theories generally are based upon opposing assumptions, motifs, models. and metaphors. Certainly selective, subjective, encapsulating factors are involved in all of them.

Very interesting in the rise of quantum theory are principles of reconciliation between quantum theory and classical physics, such as a Bohr's correspondence principle (1923), his complementarity principle (1928) and Heisenberg's uncertainty principle (1927).

BOHR'S CORRESPONDENCE PRINCIPLE

In his *correspondence principle,* Niels Bohr was aware that classical physics in its successes covered realms down to the atom as an indestructible unit. Because of subatomic minuteness, values of events within atoms escaped unnoticed until the twentieth century's focus on atomic structure. But already by the turn of the century, evidence was accumulating that the atom was not indestructible but had a physical realm of its own. The discovery of electrons, ions, X-rays, and natural radioactivity swept aside the indestructible "billiard ball" model of the atom as a prime unit of matter. A new wave of data was forcing new models: quantum theory and relativity. Atoms were found to lose or gain electrons and become ions. Metal atoms bombarded with electrons emitted X-rays. Radioactive atoms explosively emitted alpha and beta particles to transmute themselves spontaneously into other elements. The indestructible atom of classical physics was shattered. Focus had to be on the subatomic realm and on a reconciliation with the old physics. Bohr's correspondence principle held that any new physical theory, such as relativity and quantum theory, must be reconciled with the conventional physics. For example, Einstein had shown that Newtonian mechanics is a special case of Einstein's special relativity theory when the relative velocity between the observer and what he/she observes is less than about one-tenth of the speed of light (chapter 16). The smaller the relative velocity, the closer special relativity answers were to Newton's answers. And special relativity is a special case of general relativity theory in which an object traveling in four-dimensional space-time, far enough from any large masses, travels in a Euclidean straight line (chapter 17), as it also does when traveling in three-dimensional space, if observed from an inertial frame of reference.

In quantum theory, a photon striking an electron or other subatomic particle would have a large but uncertain effect upon the position or the velocity of the particle, in accordance with Heisenberg's uncertainty principle. But this same photon energy value would be unnoticed on a moving automobile or other large mass in the realm of classical physics. Thus, Bohr's bridges of correspondence between classical physics and relativity or quantum theory were valid.

We have already seen that Einstein's Equation 18.1 for the energy $E = hf$ of a photon (quantum for light) offers energy discreteness through Planck's constant h with the continuity of electromagnetic spectrum frequency f. Such contradictory opposites of discreteness and unbroken continuity in the equation suggest duality in light or other electromagnetic radiations themselves. The idea of duality[1] affords further reconciliation between quantum and classical models of energy, at least until a more all-embracing conceptual framework presents itself.

The Compton effect (1927)* provided further experimental evidence of energy-matter interaction when colliding X-rays and other quanta transferred some of their energies to recoiling electrons.

BOHR'S COMPLEMENTARITY PRINCIPLE

Bohr's *complementarity principle* (1928) recognizes the wave-particle duality of phenomena at subatomic and atomic levels. He acknowledged Heisenberg's uncertainty principle that it is operationally impossible to observe with accuracy both wave and particle aspects simultaneously. Therefore, different experimental apparatuses, e.g., diffraction gratings and photoelectric cells, reveal different aspects of light. Together the two aspects offer a more complete picture of energy at this subatomic level than either aspect alone. Bohr also acknowledged in his complementarity principle that not only light but electrons and other subatomic material particles would show wave-particle duality as determined by the particular experimental arrangement. Thus, matter has wave-like characteristics and energy has properties of matter. This should be more easily observed at the level of the minute, at subatomic or atomic levels.

In respect to matter-waves, Bohr in his complementarity principle was acknowledging ideas of the inspired French physicist Louis de Broglie (1892–1987), who, four years before, in 1924, had been the first to predict explicitly that not only energy particles but every material particle should be identified with a wave. After all, with his $E = mc^2$, Einstein in 1905 had pro-

---

*Named for Arthur H. Compton (see Fig. 18.2). Compton substituted, in energy-matter interaction, X-rays for ultraviolet light and was able to obtain higher energy electrons emitted from the metal plate as compared to what ultraviolet photons would do.

posed that matter and energy were different aspects of the same thing. If a photon as energy showed material impacts upon an electron, then why shouldn't an electron, a building block of matter, be associated with a wave nature of its own? And why wouldn't all matter have wave properties, eventually called matter-waves or de Broglie waves, if matter and energy are but different aspects of the same thing? If a photon as energy showed material impacts upon an electron, then why shouldn't an electron, a building block of matter, be associated with a wave nature of its own? Certainly a Hegelian polarity of thesis, antithesis. And why, by Einstein, wouldn't all matter have wave properties, eventually called matter-waves or de Broglie waves, if matter and energy are different aspects of the same thing? In 1925, electron waves were actually experimentally discovered by C. J. Davisson and L. H. Germer. They found that an electron beam reflected from a metallic crystal showed diffraction patterns like those of X-rays and other electromagnetic waves. Remember that electrons are subparticles of matter. Thus, matter had waves as predicted by de Broglie. Matter waves of large material objects have been difficult to detect because their calculated wavelengths are much smaller than that of an electron. Yet, while slow electrons at a velocity of 100 cm/sec have a de Broglie wavelength of 0.0727 cm, golf balls of 45.9 cm at velocities of about 2500 cm/sec have been found to have a de Broglie wavelength of 5.71×10^{32} cm. Meanwhile, the extremely powerful electron microscope has also been developed to provide magnification of over a million times, far more than any previous microscope.

CONCLUSION

Is the uncertainty in Heisenberg's principle to be understood as an uncertainty within external nature itself, in the limitations of our present knowledge of nature or as a product of knowledge-seeking man in interaction with nature? Heisenberg himself, in the development of his uncertainty principle, has said: "The object of research is no longer nature itself, but rather nature exposed to man's questioning and to the extent man also meets himself." To me, the answer lies in interaction that involves outside sources (natural events), signals and sight for observation, thought, imagination, and testing. The interaction so far involves man's intervention in the knowledge process, even at the stage of observation and measurement.

NOTE

1. *Duality* here pertains to the inclusion of opposing or contradictory characteristics of an aspect of nature: In this case it is the continuity of waves and the discontinuity of photons as con-

tradictory aspects of light. Among considerations in understanding duality in ideas of nature are these:

a. A basic polarity or paradoxical character exists in nature itself as expressed in the Chinese concepts of *yin* and *yang*, positive and negative "harmonious entities."

b. The contradictions are different aspects of the same thing often found individually under different circumstances or in interactions with different apparatus as light experiments with diffraction gratings showing wave evidence or with photoelectric cells showing particle evidence.

c. Parts of a larger picture are to be discerned later when more knowledge is available. Confidence consists here of duality forming unity with more knowledge.

d. A Hegelian thesis, antithesis, and synthesis emerges in a developmental sense through time and process rather than in a cross-sectional sense as in consideration in *a* above. That is, in consideration of *a*, the whole is a composite of two separate parts, each of which keeps its identity at any given time. In Hegel's dialectical synthesis, the interaction of parts leads to a qualitative newness of the whole, the synthesis, such as when hydrogen gas combines with oxygen to form water, a new substance.

SUGGESTION FOR FURTHER READING

Schroedinger, Erwin C. *Science, Theory and Man.* New York: Dover, 1957.

19

Causality versus Chance in Nature and Knowledge

> Predictions are based on estimates of probabilities of what will happen as a result of the interaction of a large number of independent variables. [In respect to individuals] when we predict what a certain individual will do, we are predicting the probable outcome of all the various motivations and counter-motivations he experiences.
> Maurice Cornforth, *The Open Philosophy and the Open Society,* 1968

Our purpose in this chapter is primarily to pose basic questions regarding causality or chance in nature itself or, rather, in scientific depictions of nature. Is our universe predesigned, a result of strict cause and effect, or a matter of chance? Again, can a scientist remain completely detached while observing a system of which he and his experimental apparatus are a part? It would seem that our comparative ignorance in a dynamic universe guarantees a permanent revolution in science as we acquire knowledge. In the words of Morris Kline,

> Is there a law and order in this universe or is its behavior merely the working of chance and caprice? Will the Earth and other planets continue their motions around the sun or will some unknown body, coming from great distances, rush through our planetary system and alter the course of every planet? Cannot the sun some day explode, as other suns are doing daily, and burn us all to a crisp? Was man deliberately planted on a planet especially prepared for his existence or is he merely an insignificant concomitant of accidental cosmic circumstances?[1]

A quantity of hydrogen sulfide gas, with the foul smell of rotten eggs, is placed in a corner of a large room. Since gas molecules, in accordance with Brownian motion theory, are in random motion, the path of an individual molecule cannot be predicted. And yet, a group rate of diffusion of the hydrogen-sulfide molecules has been determined; therefore, the time it takes for the odor to reach another corner of the room can be predicted. Here, certainly, is an example of order arising out of the chaos in molecular random motion, especially if we consider that all ordinary matter consists of atoms or molecules in random vibrations, whether in gas, liquid, or solid states.

In chapter 18, we saw that at the subatomic level, a limited amount of uncertainty had to be acknowledged, for example, in the interactions between photons of light necessary for observation and electrons observed. It was not possible to obtain exact measurements simultaneously for both the position of the electron and its velocity because of the intervening impact of the light photons upon the electron. This momentous insight of Werner Heisenberg led to further evidence in the development of quantum theory in terms of probability, a statistical group concept.

On the bases of randomness of molecular motion and uncertainty in interaction at submicroscopic levels, we find conceptual foundations for a universe of chance. From these foundations emerges order in the form of statistical laws of probability. Statistical laws are a perspective that enables reasoning and testing from observations not covered by classical causal laws.

The deterministic perspectives of classical physics are expressed in the famous words of Pierre Laplace which serve as an epigraph to chapter 18. An inference here is that while we are a long way from having all the necessary knowledge, we remain confident that with full knowledge, we will find the universe a strictly deterministic one of cause and effect, or possibly a designed one. Of course, a pantheist could say that since God is nature at work, nature didn't need a beginning. Something may have preceded the Big Bang as in a pulsating universe.

Thus, as we have seen also in the previous chapter, we have a modern problem of perspective in a scientific search for order or unity in nature. Like the contradictory concepts of *waves* and *particles, causality* and *chance* may be dual aspects of a larger picture (not yet fully determined) for unifying knowledge. The problem may be one of extrapolation from macroscopic levels of experience in our everyday world to this submicroscopic realm of the atom. The causality concept had its exciting successes in the macroscopic or intermediate Newtonian realm. But to apply the causality concept in its original connotation to the subatomic realm may be like applying the concept of heat energy in kinetic-molecular theory—with its focus on *molecular* motion—to subatomic realms *below* molecules. Such extrapolation simply does not apply to the new realm. Prediction is the heart of science. Previous predictions were based upon causal laws of certainty; in the new submicroscopic realm, however, statistical laws of probability are more appropriate for prediction, at least so far.

One cannot talk of statistical laws in terms of classical laws. Laws of cause and effect predicting events and focused on individual isolated entities are not the same as laws of chance involving statistical averages of randomly moving groups of molecules. As James R. Newman expressed it, "What appears to be predictable events on the macroscopic level are strictly statistical averages of a vast aggregate of irregular unpredictable events on the microscopic level."[2] Gas pressure within a blown-up balloon is an example. This is shown by the random motion on the walls of the balloon.

Thus, the meaning of causality for classical, dynamic laws is different from any possible meaning for statistical laws. In fact, *correlation* is probably a better word than causality for general statistical formulations, even for those supporting Heisenberg's uncertainty principle.

Quantum mechanics, with its uncertainty principle, still suffers at this point from a comparative weakness. Although it offers successful *group* prediction—as in rates of molecular diffusion for different freed gases—its predictability for individual objects within a group is inadequate, to say the least. It limits possible information. At this point, it would seem that each perspective should be used for whatever information each can add just as contradictory wave theories and particle theories of light have provided complementary information and insights about light. Besides quantum mechanics flowed from Bohr's correspondence principle: Under particular circumstances, classical physics offers good approximations for quantum mechanics. Two examples of this correspondence, already discussed in chapters 16 and 17, are Newton's mechanics as a special case of Einstein's special relativity theory and Einstein's special theory as a special case of general relativity. In chapter 16, we also saw that with Einstein's $E = mc^2$, matter and energy are interconvertible and that the photon or quantum of light energy has material aspects of mass and impact. Light's impact on metal electrons leads to Heisenberg's uncertainty principle and eventually quantum mechanics.

Alexei Matveyev[3] makes an excellent point when he illustrates that no causality may exist in a chance connection of an auto accident. He emphasizes that an accident between two cars involves two chains of events. Each car has its own chain of events in which causality can be shown. But no causality exists in the *chance connection* of the accident. That is, a change in the chain of events for one car need *not* affect the independent chain of events for the other. Although in that case the accident need not occur, there would be no physical law of causality *between the two chains of events to predict an accident,* and no question of indeterminacy need be involved. In respect to connections between the two chains of events themselves, both causality and indeterminacy would be meaningless.

David Bohm,[4] backed by Louis de Broglie, has opened up in recent years an ingeniously plausible argument for reinforcing determinism in the present causality versus chance debate. With the breakdown of the atom have come subatomic levels of innumerable particles , radiations, photons, matter-waves, and so on. Who is to say that today's quarks are the indivisible lowest unit of reality any more than atoms themselves were a century ago? Bohm proposes the possibility that phenomena and principles at a subquantum-mechanical level, as yet undiscovered, could remove the uncertainty in the uncertainty principle at the higher level of electrons and photons. That is, this as yet undiscovered underlying level could provide necessary information for precision at the higher electron-photon level. Bohm articulates the plausibility of an imaginary sub-basement model that in detail could serve his deterministic outlook.

R. A. Fisher[5] has emphasized that no isolated systems exist in nature and, therefore, no *single* variables operating in classical, deterministic experimentation can give precise results. With his famous *analysis of variance and covariance,* he developed pioneering statistical designs to allow for more than one variable to operate in scientific and life situations. Conclusions would be closer to nature or actual experimental situations even though in terms of "levels of significance," a statistical term in line with probability rather than with strict determinism or causality.

In respect to the human dimension in all this, we must refer the reader to chapters 10 and 11.

NOTES

1. Morris Kline, *Mathematics in Western Culture* (New York: Oxford University Press, 1953), ch. 24: "Our Disorderly Universe."
2. James R. Newman, "Review of David Bohm, *Causality and Chance in Modern Physics,*" *Scientific American,* July 1959.
3. Alexei Matveyev, *UNESCO Colloquium, Science and Synthesis* (New York: Springer-Verlag, 1971), p. 141.
4. David Bohm, *Causality and Chance in Modern Physics* (New York: Harper, 1961). This is an ingenious attempt to reconcile determinism with twentieth-century developments in quantum theory uncertainty.
5. R. A. Fisher, *Statistical Methods for Research Workers* (London: Oliver & Boyd, 1925). Here is a pioneering work in providing ingenious statistical methods for more than one variable operating in scientific and life situations.

SUGGESTIONS FOR FURTHER READING

Bohr, Niels. *Atomic Physics and Human Knowledge.* New York: Wiley, 1958.
Hanson, Norwood Russell. "Copenhagen Interpretation of Quantum Theory." *American Journal of Physics,* January 1959.
Johnson, Palmer O. *Statistical Methods in Research.* New York: Prentice Hall, 1949. A university text by an enthusiastic disciple of Fisher for advanced statistical students.
Kline, Morris. *Mathematics in Western Culture.* New York: Oxford University Press, 1953. Chapter 24: "Our Disorderly Universe: The Statistical View of Nature." This is an excel-

lent chapter for introducing issues of causality versus chance in nature. The chapter, an imaginary debate between a Mr. Determinism and a Mr. Probability, is a clear presentation of the issue from the statistical, probability point of view.

Schilp, Paul A. *Albert Einstein: Philosopher-Scientist*. New York: Tudor, 1951. Presents biographical materials and prominent debates on causality versus chance based upon modern physics and philosophy.

Schroedinger, Erwin C. *Science, Theory and Man*. New York: Dover, 1957. A lucid collection of essays on philosophical and humanistic problems in contemporary physical science by an outstanding contributor to quantum theory.

Walker, Helen M. *Elementary Statistical Methods*. New York: Henry Holt, 1947. An excellent introduction to Fisher's *Analysis of Variance and Covariance*.

For relevant side interest, see Philip D. Zelazo, University of Toronto, and Douglas Frye, New York University, *Science News*, July 17, 1993, p. 42: both psychologists contend, based upon cognitive research with youngsters from three to about seven years of age, that "4-year-olds have an emergent general ability to reason first from one perspective and then from another, incompatible perspective." See also Alison Gopnik, a University of California, Berkeley, psychologist, who further contends, based upon her experiments, that "The same mental capacities that children use to understand the mind have been applied to science by adults. It's not that children are little scientists, but scientists are big children." She reportedly "chuckled at the implication." *Science News*, July 17, 1993, p. 40.

Part Four
Interactive Systems within Systems

20

Prediction as Interaction: Cybernetic Feedback

Man's advantage over the rest of nature is that he has the physiological and hence the intellectual equipment to adapt himself to radical changes in his environment.... [E]ffective behavior must be informed by some sort of feedback process, telling it whether it has equalled its goal or fallen short.
                                   Norbert Wiener, *The Human Use of Human Beings,* 1950

The three twentieth-century revolutions in physics have been relativity, quantum theory, and cybernetics. The technological concomitant of cybernetics, of course, is the astounding computer development now in full bloom.

Webster's Collegiate Dictionary (10th ed., 1993) defines cybernetics as "the science of communication and control theory that is concerned especially with the comparative study of automatic control systems as the nervous system and brain and mechanical-electrical communication systems." B. M. Kedrov quite concisely defines cybernetics as "self-regulating mechanisms of [control] systems and of human activity."[1] Norbert Wiener (1894–1964), the brilliant U.S. mathematician, was founder of cybernetics as a contemporary science. He invented the term "cybernetics" from the Greek word *kupernetes* or "steersman," the same word from which we derive our word "governor."[2] Wiener's great pioneering masterwork, *Cybernetics: Control and Communication in the Animal and the Machine,*[3] first published in 1948, brought his innovating discipline to scientific and public attention. He considered cybernetics to be an interdisciplinary scientific theory relating neural networks, computers, learning theory, communication theory, servomechanics, and automatic control systems.

The title itself of Wiener's well-known popular version of his ideas, *The Human Use of Human Beings: Cybernetics and Society* (1950), indicates his interdisciplinary bent in science. Certainly significant for our purposes here is Wiener's general feedback concept and his emphasis on the self-regulation of

313

mechanical-electrical systems, of living systems, of scientific investigation, and of human activity in general.

As a starter, the following simplified schematic briefly generalizes cybernetic feedback for learning and adjustment:

INFORMATION → COMMUNICATION → FEEDBACK → CONTROL modification
from outside of   to responsive   from device
a system   system device

Let us consider the following illustrations:

I. Feedback Systems as Automatic Processes in Machines

(1)   Cause brings about an effect (1). Effect (1) becomes cause (2) creating an effect (2). (See thermostat example and Figs. 20.1 and 20.2 below.)
Cause (1) → effect (1). Effect (1) becomes the cause (2) → effect (2).
Effect (2) eliminates the original cause (1).
System becomes a cycle of cause and effect, and self-regulatory.

Example of Feedback Thermostatic Action of a Furnace

Cause (1) cold   drop   thermo-   electri- increase   effect (1)
          outside→room→stat→   switch→cal→   in→
          air   temp. metals on   energy   heat

Effect (1) increase rise in thermo-
becomes in→   room→stat→   switch→no→   decrease→ effect (2)
Cause (2) heat   temp. metals off   energy   in
                                                      heat

Two different metal strips as iron and brass when riveted or welded together form a compound bar. When heated, the bar will curl or bend (Fig. 20.1) because the brass expands more than the iron.

The compound bar becomes a *thermostat* when it has a free end arranged as in Fig. 20.2 to close a proper electric circuit and thereby control temperature of an enclosed space.

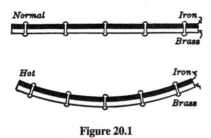

Figure 20.1

Effect of heating a compound metal bar.

2. Human Application: Double Thermostatic Action of the Body in Temperature Regulation.

Cause (1)	cold → weather	stimulus→ on cold sensory nerve endings	sensory→ nerve electric impulse	thermo-→ stat: hypo- thalmus	motor→ nerve electrical impulse	effect (1) a. muscle contraction: heat generated

Effect (1) a, b, c becomes cause (2) temperature rises, original stimulation void.

effect (1)

a. muscle contraction: heat generated

b. capillaries contract: less surface blood; less heat loss.

c. Sweat glands inhibited: less evaporation; less cooling.

Note: *Heat* effect on nerve endings follows a similar pattern.

3. Scientific Inquiry and Feedback

Fig. 11.A and B emphasized scientific inquiry as interaction with nature. Fig. 11.A diagrammatically depicted the self-correcting investigation process and Fig. 11.B components of (this) investigation. Details covered at considerable length in chapter 11 do not need repeating here. We will merely repeat Fig. 11.A and B as Fig. 20.3A and B, and list the processes and components in sequence: natural events (facts), signals (from facts), perception (data), patterning (perceptual and conceptual), projecting (back on nature), predicting (hypothesis testing), feedback and corrections, perpetual reorganization of data, and perspective: tentative systems within systems.

Once again, interaction is involved between scientists and nature in the

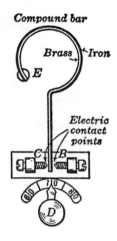

Figure 20.2

Thermostat for controlling room temperature.

investigation of nature. Scientists form systems with what they observe, and feedback takes place within the interactive process. Feedback corrections can be either of observations or of ideas.

4. Psychosomatic Difficulties

Causes

Stimulation by problems, worry, fear, conflict, anxiety, frustration, etc.

Effects

    a. The body, through the activated nervous system, affects glandular secretions, dilation or constriction of the blood vessels, amounts of blood sugars, etc.

    b. The system becomes charged with tensions and energies for action:
        IF energies result in action of some kind, the original cause may be removed.
        IF the removing action is not possible, the nerve impulse will go into some internal channel such as the digestive system, causing ulcers, colitis, blood vessel spasms, blood pressure problems, etc.

    c. The system affected will be determined by the automatic nervous system and the subconscious.

5. Social Implications

Wars, racial discrimination, social insecurity, etc., result from inept, hasty-feedback social systems and responses.

Entropy versus System Building

Entropy, sometimes referred as "heat death of the universe," is defined in the tenth edition of the *Merriam-Webster Collegiate Dictionary* (1993) as a "thermodynamic measure of the amount of energy *unavailable* for useful work in a system undergoing change" (italics added) as, for example, loss of energy due to friction. This definition leads to another: "a measure of the degree of disorder in a substance or system; entropy always increases and available energy diminishes in a closed system, as the universe." Also, "in information theory and computer science, [entropy is] a measure of the information content of a message evaluated as to its uncertainty."

The universe itself in its entirety may be considered an isolated system in entropy. But there is also an opposing universal tendency of increasing organization and complexity. Within our sun and the other stars, the simplest atoms, hydrogen, fuse into helium and more complex elements, and so on until there are built up in the stars, elements of chemistry's Periodic Table. Atoms of various elements combine to form molecules or compounds of increased complexity and organization, even to the extent of single-celled plants or animals evolving. Simple forms of life developed into more complex forms with Charles Darwin's explanation of individual variations and natural selection, and with Hugo de Vries's mutations. In photosynthesis, green plants use the sun's rays to form starch from water and carbon dioxide. And ultimately complex systems within systems integrate into the marvel of human awareness of self and surroundings. Certainly life at any level has tendencies to survive, to multiply, and to organize.

Thus, at present, with Wiener's cybernetic and computer theories, and with general scientific system building, we find a *negative entropy*, an organizational buildup in antitheses to entropic breaking-down systems—unless, of course, we destroy ourselves by the dangerous gap now existing between technological advances and social intelligence. Are we helping "heat death" or chaos to win here on earth? As Norbert Wiener expressed it, "We have modified our environment so radically that we must now modify ourselves in order to exist in this new environment. We can no longer live in the old one."[4] Do we humans have what it takes for social reorganization in line with our technological advances?

Fig. 20.3
Scientific Inquiry as Interaction with Nature

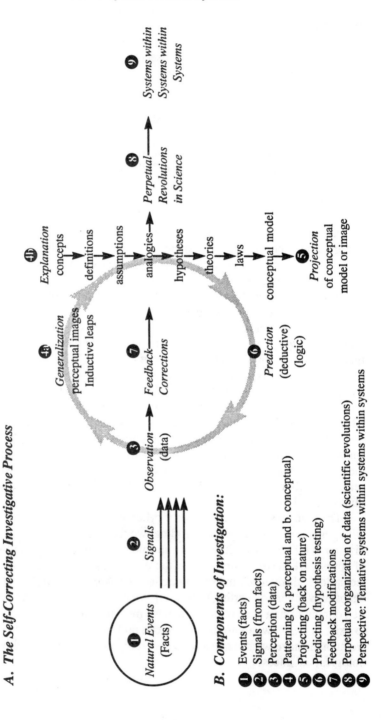

A. *The Self-Correcting Investigative Process*

B. *Components of Investigation:*

❶ Events (facts)
❷ Signals (from facts)
❸ Perception (data)
❹ Patterning (a. perceptual and b. conceptual)
❺ Projecting (back on nature)
❻ Predicting (hypothesis testing)
❼ Feedback modifications
❽ Perpetual reorganization of data (scientific revolutions)
❾ Perspective: Tentative systems within systems within systems

Meanwhile, we repeat Wiener's words: "Semantically significant information in the machine as well as in man is information which gets through to an activating mechanism in the system that receives it, despite man's and/or nature's attempts to subvert it. From the point of view of cybernetics, semantics defines the extent of meaning and controls its loss in a communications system."[5]

Wiener puts us further on our guard with these words:

> Pere Dubarle has called the attention of the scientist to the growing military and political mechanizations of the world as a great superhuman apparatus working on cybernetic principles. In order to avoid the manifold dangers of this, both external and internal, he is quite right in his emphasis on the need for the cultural anthropologist and the philosopher. In other words, we must know as scientists what man's nature is and what his built-in purposes are, even when we must wield this knowledge as soldiers and as statesmen; and we must know why we wish to control him.[6]

NOTES

1. B. M. Kedrov, *Science and Synthesis* (New York: Springer-Verlag, 1971), p. 151.
2. Norbert Wiener, *The Human Use of Human Beings: Cybernetics and Society* (New York: Avon, 1967), p. 23; first published in 1950 for the intelligent, nonscientific reader.
3. Norbert Wiener, *Cybernetics: Control and Communication in the Animal and the Machine* (Boston: MIT Press, 1948), which first brought his new innovating discipline to scientific and public attention.
4. Wiener, *The Human Use of Human Beings,* pp. 249–50.
5. Ibid., p. 128.
6. Ibid., pp. 249–50.

SUGGESTIONS FOR FURTHER READING

Crasson, Frederik J., and Kenneth S. Sayre, eds. *Philosophy and Cybernetics.* New York: Simon and Schuster, 1967.
De Vries, Hugo. *Die Mutationstheorie (Mutation Theory),* 1901. In this work de Vries coined the term "mutation" as a sudden change in the evolutionary process that resulted in a new biological species. He then investigated experimentally for about eight years. A modified version of this idea of mutation still persists in evolutionary theory.

21

The Human Microcosm: Interactive Worlds within Worlds

The web is an image for the whole of creation, symbolizing the interconnection and interdependence of all its constituent parts. In this web, human
beings are linked with animals, plants, land, water, rocks, birds. . . . Whatever
happens to one part is felt in some way in all other parts. Like a web, creation
is intricate and beautiful. And like a web, it is fragile and easily damaged, its
integrity or wholeness easily destroyed.

> Kevin W. Kelley, *The Home Planet,* 1988[1]

In the world of life, nature, and things, there is no artificial separation of
physics, chemistry, geology, astronomy, or biology. Light and electricity have
characteristics of both matter and energy; modern physics and chemistry share
common ground in atomic science. The human body incorporates almost all
the sciences. It would seem that understanding nature and ourselves should
reflect the integration of interrelated parts that actually exists in the world and
in life.

If we humans are a conscious part of nature *interacting* with the rest of
nature in science and otherwise, such interaction may afford insights into both
nature and ourselves not otherwise possible. The materials just below are
meant to suggest an approach to a study of man based on interactive concepts
derived from the natural sciences. In this we take a cue from the idea of the
ancient Greeks that each of us is a microcosm, or miniature universe. But we
emphasize here that as microcosms we are not island universes; we interact
with other microcosms and are parts of a vast macrocosm. We inhabit surroundings that shape us and that are shaped by us. We incorporate almost all
the sciences, the social sciences, and the humanities, too.

To illustrate and to provide details and further thought, let me draw from
a lengthy article I wrote some years ago on integration in college science
courses for general education.[2] Science is primarily a matter of asking ques-

tions of nature which is arranged, hopefully, to receive answers in feedback fashion. The following extracts therefore suggest this in a series of key questions and suggested study projects. Our purpose in this chapter is not to repeat classroom plans in science general education already published. Rather, it is to concisely review sufficient details for a general "wholistic" outlook of "worlds within worlds or systems at increasing levels of complexity."

We start with the following three propositions:

1. The human personality includes energy mechanisms adapted for change in a world of change at levels that may be designated as subatomic, chemical, physical, physiological, biological, psychological, sociocultural, and cosmological.

2. The activities at these various levels are interrelated. Variations in activity at any level affect activities at other levels.

3. The internal activities, interactions, and changes of the human microcosm can be understood and often predicted in terms of basic laws within and between the various levels of organization, including the sociocultural.

These three propositions may be reinforced by such generalizations as the following:

1. The matter making up our bodies has the same electrical basis and is subject to the same subatomic, electrical, and chemical laws as other objects around us.

2. The body in its locomotive, temperature-regulation, and sensory apparatus is equipped for dynamic physical relationships with a changing environment and is subject to universal laws of mechanics, thermodynamics, light, and sound.

3. The body, through its own composition; its cellular activity; and its digestive, respiratory, and glandular systems, performs chemical transformations for its own needs.

4. The cells, tissues, organs, nervous system, and brain with their specialized functions form a complex integration of activity for the well-being of the human organism in its surroundings.

5. Associated with the various activities above, biological, neurological, psychological, and sociocultural mechanisms condition the goal-seeking interactions of the human personality and society with its environment; humans, in turn, affect the environment by their interactions.

6. The scientist, in his or her observations, explanations, predictions, and technological applications, is a case in point of organized feedback in inter-

action with his or her surroundings, immediate and cosmological. Developmentally, there was ancient biblical cosmology, then astrology and alchemy as "fossil sciences," then modern astronomy with its dynamic universe expanding from a "Big Bang." All these reveal cosmological interactions in a developing survival/cultural search for order that hopefully helps create order.

7. Thus, the human organism is considered as a microcosm, as an integrated complex of activity at different interrelated levels—physical, chemical, physiological, biological, psychological, and sociocultural—reacting to an ever changing environment, with internal changes at one level affecting changes at other levels.

My article was written independently of Ludwig von Bertalanffy, whom I discuss in chapter 22. I first became familiar with von Bertalanffy's book *General System Theory* by reading his article written for *Main Currents* in 1955. My article and von Bertalanffy's book serve different purposes in appreciation of integrative education.

THE HUMAN MICROCOSM: INTERRELATED LEVELS OF ACTIVITY

Our suggested series of questions and relevant projects ranges from simpler activities on subatomic levels to ever more complex aspects of our psychological and social existence. The "simpler" section immediately below evolves into more and more interrelatedness in succeeding sections and culminates hopefully in concepts of systems within systems or worlds within worlds. The outline form makes for conciseness.

I. Activity on the *Subatomic Level* of Your Body

A. Key questions

1. What changes take place in your body on the subatomic level?
2. What external subatomic dangers does your body face?
3. What can you do for control?

B. Possible topics for study

1. Why should you not stand under a tree in a thunderstorm?
   a. What is lightning?
      (1) What were ancient ideas of lightning?
      (2) What was Benjamin Franklin's experiment with a kite? Why was it dangerous?

       (3) Demonstrate electrical charges and their attraction and repulsion with such devices as pith balls, electroscopes, glass and amber rods, silk cloth, and fur.

   b. Where are the electrons within your body?

       (1) What is the normal arrangement of electrons in the atoms within you?

       (2) Are electrons matter, energy, or both? What evidence is there?

       (3) Why, after you walk across a carpet, will a spark jump between you and a metal object toward which you are reaching?

       (4) Why can the comb you just used in your hair pick up small bits of paper?

   c. Why will a tree be severely damaged by lightning while a flag pole on the same ground is not?

   d. What has been or can be done to protect you at home or at work against lightning?

2. Why is it dangerous for you to place a light switch or light fixture near a bathtub?

   a. What is the nature of an electric current?

   b. Look into the work of two or three of the following men in electricity: Volta, Oersted, Faraday, Maxwell, Edison, Steinmetz, or any other you may choose.

   c. Explain a parallel electrical circuit in a home with the use of switches, fuses, etc. What is the benefit of parallel circuits?

   d. Explain how a stream running downhill is like an electric current.

   e. List six good conductors and six poor conductors of electricity. Is the hard water in your bath a good conductor or a poor conductor? When can this be dangerous for you?

3. Why can a bird "get away with" perching on a live wire?

4. What are *quarks*? Where are they located within you?

5. Discovery and development of radium, X-rays, and other radiations.

   a. Look into the works of Madame Curie, Roentgen, Becquerel, or any other creative worker in radioactivity.

   b. What are some of the medical and other uses of radium and of X-rays?

   c. Describe dangers and safeguards in the use of radium and X-rays.

6. Why can it be said that nuclear fission means one world or none?

   a. What do we mean by "atom"?

       (1) Read several accounts of the early alchemists and what they were attempting.

(2) In what way and to what extent has modern science accomplished what the alchemists were attempting?

    (*a*) Where are the atoms within your body and describe their composition?

    (*b*) In what way are these atoms like solar systems?

b. What do you know about the atomic bomb?

  (1) Give a report on one of the government atomic plants operated during World War II.

  (2) What safeguards are necessary in all atomic energy plants?

  (3) Give a report bringing in clippings or illustrations on one of the two atomic bomb attacks on Japan or on one of the government tests at Bikini or in New Mexico.

c. A future atomic war.

  (1) Report on H. G. Wells's *The Shape of Things to Come.*

  (2) Investigate present ideas as to the nature of an atomic war.

  (3) What do you think is the basis for Einstein's statement that he thought there was no defense against the atomic bomb?

  (4) What do you think could be done in the matter of relationships among people and nations to reduce the possibility of another war?

II. Activity on the *Physical Level* of Your Body.

A. Key question: How is your body equipped for physical contact with your environment?

This in turn involves such considerations as

1. the human being as an integrated, dynamic energy mechanism. What are dynamic and integrative aspects of the human personality?
2. how the body is an instrument for locomotion and for work
3. how the body is equipped for temperature and heat changes within and around itself
4. how the body is provided with sensory apparatus of sight, sound, taste, and smell in its environmental contacts.

B. Projects could center about the following basic aspects of biophysics:

1. Matter in motion considerations of the body:
   a. the nature and status of matter, and the kinetic-molecular theory
   b. special properties of matter as applied to the body, its tissues, tissue functions, and structure

    c. weight, mass, volume, center of gravity, posture, structure, equilibrium, and muscle tonus of the body

    d. specific gravity and buoyancy of the body as well as changing pressures within and upon it

    e. the mechanics of the heart and lungs: blood pressure and breathing

    f. the functions of the skeleton and muscles in locomotion

    g. Newton's laws of motion: inertia, acceleration, action and reaction, as they apply to the body

    h. centrifugal, frictional, and gravitational forces for the body to overcome

    i. laws of falling bodies as they apply to vertical and horizontal motion

    j. force, work, mechanical efficiency, energy, momentum, and power of the body

    k. the six basic machines (lever, inclined plane, wedge, screw, and pulley), their effective use as tools and their significance.

2. Thermodynamic considerations of the body:

    a. the kinetic theory of heat and its application to understanding the thermodynamic considerations and processes of the body

    b. the constant temperature of the body in its balance between released chemical heat and evaporation of perspiration

    c. thermometry and its applications

    d. considerations of relative humidity, clothing, heat transmission, and expansion due to heat in reference to body comfort and proper body function

    e. safety and health considerations of the heat both of vaporization and of fusion

    f. health considerations of weather and proper clothing involving condensation and precipitation as dew, frost, clouds, fog, rain, snow, ice, melting point, and vaporization.

3. The sensory apparatus of your body:

    a. the eyes as mechanisms of sight

        (1) the nature and mechanism of light

        (2) the eyes as a receiving mechanism:

            (*a*) comparison to a camera

            (*b*) superiority of the eyes to a camera in the accommodation of eye lenses, persistence of vision, and three-dimensional images

            (*c*) optical illusions due to reflection and refraction

            (*d*) color and dispersion of light

        (3) optical devices as eye improvements and extensions:

(*a*) glasses for refractive correction of eye defects

(*b*) reflective benefits of the mirror, sextant, and periscope

(*c*) "magnification" by magnifying glasses, telescopes, microscopes, and binoculars

(*d*) the permanent records of the ordinary and moving picture camera

(4) Safety and health considerations involving your eyes:

(*a*) proper intensity of illumination involving law of inverse squares

(*b*) proper diffusion and removal of glare

(*c*) color blindness

(*d*) psychological effects of color

(*e*) highway lights and traffic signs

b. your ears as mechanisms of sound:

(1) the nature and mechanism of sound

(2) the ears as receiving mechanisms:

(*a*) comparisons to telephone receivers or loudspeakers

(*b*) the mechanism of the ear

(*c*) aural illusions as in cases involving Doppler's principle

(3) Aural devices as aids to the ears:

(*a*) hearing aids for hearing deficiencies

(*b*) magnification by megaphones, microphones, and amplifiers

(*c*) acoustical principles and considerations

(4) Safety and health considerations involving the ears:

(a) sanitation and hygiene involving, for example, removal of ear wax and foreign bodies from the ear

(*b*) intensity of sound

(*c*) psychological effects of noise

(*d*) the use of sirens

(5) the explanation and appreciation of musical instruments:

c. the senses of smell and taste for stimulation of body digestive processes and for danger signals.

III. Activity on the *Chemical Level* of Your Body

A. *Key questions*

1. What changes take place in your body on a chemical level?

2. What external assistance does your body require?
3. What dangers does your body face on this level? What effects are there upon other aspects of our lives?
4. What can we do for control?

B. *Possible projects for investigation, e.g., nutrition, digestion, and metabolism*

1. What is the role of food and water in the chemistry of our bodies?
   a. Make a list of the chemical elements that go to make up the body.
   b. Keep a list of the food you eat during the next week. Then by classifying these foods primarily according to sugars, starches, fats, proteins, minerals, vitamins, and cellulose, determine which of the necessary twelve chemical elements you are obtaining from each food.
   c. How is it possible for you to be overweight and still be suffering from malnutrition?
   d. Plan and record a week's diet that you would consider balanced and explain why you think it is balanced.
   e. Obtain from your doctor, a hospital, school dietitian, or any other expert a diet that he or she would consider balanced, and compare it, to *b* and *d.*
   f. What specific roles does each of the following play in the body: sugary, starchy, and fatty foods; proteins; minerals; the various vitamins; water; and fibers?
   g. Make a list of five foods each rich in proteins; carbohydrates; fat; calcium; vitamins A, B, C, D; and also five that are over 50 percent water.
   h. By actual chemical tests, determine what food elements are in any five of the foods you included in list *b.*
   i. Test five items in this list for acids and five for alkalies. What chemical elements do acids and alkalies contain?
   j. What is the great benefit to health in the fact that acids and alkalies neutralize each other? Mix some lemon juice and baking soda, and describe what you saw. Show what happened by a chemical equation.
   k. By consulting a calorie chart, figure your average daily caloric intake from your list in *b,* and check it against what is advisable with your daily activity.
   l. Collect dinner menus from several restaurants and analyze them from the standpoint of balanced diet and calories.
   m. Make a drawing showing the passage of food from the mouth through the rectum, indicating at the proper positions the salivary

glands, esophagus, stomach, liver, gall bladder, pancreas, duodenum, large intestines, appendix, colon, and rectum. Describe the part each of these plays in the breakdown of food.

n. What is now being done with radioactive particles to trace the distribution of food elements throughout the body?

o. What diets are characteristic of the following diseases and explain in each case the purpose of that particular diet: diabetes, kidney or Bright's disease, gout, anemia, and ulcers.

p. Make a table showing what may result from a serious deficiency of each of the following in the diet: carbohydrates, proteins, such minerals as calcium and iodine, the various vitamins, fibrous material, and water.

q. What condiments have you eaten during the last week? What are the dangers of condiments? What purposes do they serve?

r. Develop cultures of molds and bacterial decay on pieces of bread, meat, or fish. Also devise and describe an experiment showing the keeping qualities of pasteurized and unpasteurized milk. Name and describe five modern practices that have reduced food poisoning.

s. How does the sun "feed" us with one of the food elements through our skin?

t. Obtain from the United States government printing offices a copy of the present Federal Food, Drug and Cosmetic Act and indicate, by written report, the protection that each provision gives.

u. Find newspaper or magazine clippings pertaining to food or health advertising and evaluate each one.

v. Go to your nearest canning factory, health food store, or neighbor owning a vegetable juicer and ask to see how the juices are made. How would you explain this easy liquifying of vegetables and the small residue?

w. What is *metabolism*? Its importance? Its mechanism?

x. How does metabolism affect personality?

y. What differences in personality may be predicted by metabolic differences?

z. What percent of the body is water? In what ways does the body get its water? What are the various functions of water in the body? What precautions are necessary for drinking water?

aa. How does your system dispose of food elements that it cannot oxidize? For example, trace the path that a nitrogen particle may take after absorption from the next piece of meat that you eat.

bb. Enumerate and describe at least three ways in which your system disposes of the waste products in your body.

IV. Activity on the *Physiological Level* of Your Body

The following, as in all other phases of this section, is intended merely to be suggestive:

A. *Key questions*

1. What activities go on in our bodies on a physiological basis?
2. What disturbances of an internal or external nature are there in connection with these functions?
3. How can you help to overcome these disturbances and hazards?
4. What advantages do body cells, and the body itself, have over other energy machines?
5. Of what importance is protoplasm in understanding yourself as a being?

B. *Activities*

1. Microscope demonstration as
   a. the basic physiological unit, the cell
   b. types of cells and tissues according to function, such as blood cells, skin or hair, etc.

2. Chart the various physiological processes of the body, indicate the organs involved, and show how the tissues vary according to function.

3. Demonstrate by charts, models, and skeletons the locations and functions of different parts of the body.

4. List ten common diseases caused by germs or viruses, including all those you can remember having had. In chart form, indicate:
   a. the cause of the disease
   b. ways in which bacteria enter the body
   c. the tissues, areas, and functions involved
   d. ways in which the body defends itself
   e. what we can do in advance to prevent such diseases
   f. the nature of outside assistance to overcome such diseases.

5. List six other illnesses not caused by germs or viruses, including all those you have had. In chart form, indicate:
   a. the cause of the illness
   b. the tissues, areas, and functions involved
   c. the way in which the body defends itself

    d. what we can do in advance to prevent such illnesses

    e. the nature of outside assistance against such illnesses

6. Prepare a written or oral report on the history and significant statistics of one disease or illness listed in 4 or 5.

7. Make an outline of the sole of your shoe. Then, outline your bare foot as you stand on the first outline. Are you wearing proper shoes? Do you wear tight fitting clothes of any kind?

8. Chart as long a list as you can of physical injuries that the body or its parts can experience other than through bacteria or illness, and make a list of do's and don'ts that would prevent such injuries.

9. What are mastoids? Why are they particularly dangerous?

10. Why is the nervous system like a telephone system? Report on the work of Sister Kenny. What precautions can you take against infantile paralysis?

11. What is the function of pain or headaches? Why is it ordinarily a poor policy to develop the habit of taking aspirin?

12. What are the harmful and beneficial effects of alcohol on a chemical, physical, physiological, and psychological basis? Visit or get information from such an organization as Alcoholics Anonymous.

13. What are the four blood types? Which type are you? Investigate the R_H factor.

14. How are diseases prevented by vaccination, antitoxin, or serum immunity?

    a. How do vaccinations prevent diseases?

    b. What is the principle underlying the treatment for diphtheria, the Schick test, the Dick test? Describe the vaccines against smallpox, typhoid, boils.

    c. What is the preparation, use, and value of tetanus serum? How does tetanus develop from rusty nails or gunshot wounds?

15. What is meant by natural immunity?

16. Discuss the function and origin of the white blood cells.

17. What is the relationship between allergies, hay fever, asthma, and migraine headaches? Discuss causes and treatments of them. How much is known about prevention of allergies?

18. How can we prevent disease by controlling insect and rodent carriers?

a. Discuss the part each of the following plays in disease and what can be done to control them: flies, mosquitoes, fleas, lice, bedbugs, ticks, tapeworm, and rats.
b. Observe the foot of a fly under a microscope. Also expose a piece of meat to flies: Watch and describe the stages in the development of flies such as maggots, pupae, etc.

19. Make as large a list as you can of the ways in which bacteria can get into your body and give one illustration of a disease for each method. What precautions can you take in minimizing the entrance of bacteria into your system?

20. Look into the work of several of the following: Louis Pasteur, Joseph Lister, Robert Koch, Edward Jenner, Theobold Smith, Walter Reed, and Paul Ehrlich.

V. Biological, Psychological, and Social Considerations of Yourself

A. *Key questions*

1. What are the various aspects of personality and behavior that go to make you, you?
2. What biological mechanisms are involved in the shaping and perpetuation of yourself?
3. What personality differences or disturbances arise from differences or disturbances on lower levels? E.g., how do glands and the nervous systems perform their dynamic, integrating roles in physiological, mental, and emotional processes?
4. What are visceral, activity, sensory, and emotional drives and how are they related to behavior?
5. How does your environment shape you?

B. *Activities*

1. What are the various aspects of personality and behavior that go to make you, you?
   a. List the various factors that go to make up your personality.
   b. Take someone whom you consider to have a great deal of personality and explain just what it is that seems to make up that personality.
   c. Discuss the following as instruments for change, for resistance to change, or for both: instincts, reflexes, habits, intelligence, and language.

d. What is the function of fear? Give three examples where fear was of benefit to you; three examples where it did you harm.

e. Consider to what extent worry, fear, and insecurity play an all-important role in the world situation today.

f. List five things you do as reflexes, instincts, habit, and reflective thinking.

g. What are the different functions in desirable behavior that reflective and critical thinking as well as habit can play?

h. Cite three examples of where during the last week you used reflective thinking in solving a problem, three examples where you used critical thinking, and three examples where you should have applied critical thinking but did not.

i. Write a report on William James's book "Habit." Give ten examples of how you resist change. Discuss the good and bad aspects of this resistance. Give ten examples of how society resists change. Discuss the good and bad aspects of that.

j. What is intelligence according to such different schools of thought as behaviorism, Freudianism and Gestalt psychology, etc? What do you think intelligence is?

k. To what extent do intelligence tests measure intelligence rather than ability at school? To what extent does intelligence involve adjustment to change?

l. Have you ever walked in your sleep? How do you know? Explain sleepwalking.

m. Under what conditions could you be hypnotized and what does this reveal about a person's mental and emotional processes?

n. Comment upon Sigmund Freud's analysis of dreams. What did Freud mean by the subconscious mind?

o. List the ways in which you consider your emotions to be an asset. When are they disadvantageous?

p. What is a psychosomatic illness? Illustrate. How can this be explained in terms of body-mind-emotion interrelationships?

q. You and change:

(1) List five ways in which you are physically different now than you were five years ago. List five ways in which you will probably be different five years from now.

(2) List five interests that you have today that you did not have five years ago. In what way will these interests change five years from now? Why did and will these changes take place?

(3) List five things about which you changed your mind during the last six months. Do you expect to have the same opinions on

everything six months from now that you have today? Will these changes be due to the fact that you have changed, that things around you have changed, or both?

(4) What is happiness?
   (a) List what you consider you need to give you peace of mind and happiness.
   (b) Would it have been the same things five years ago?
   (c) Will the above be the same five years from now?

2. What biological mechanisms are involved in the shaping and perpetuations of yourself?
   a. What is the Mendelian Law?
      (1) Describe a characteristic that you think may run in some family that you know.
      (2) Explain what part genes, chromosomes, or hormones may have played in the transmission of that characteristic.
      (3) Examine sex chromosome slides under a microscope.
      (4) In a home lab experiment, illustrate the Mendelian law of distribution of characteristics by picking blindfolded from 400 marked and unmarked toothpicks that have been equally distributed into two piles, and by recording your end results.
      (5) If both parents have brown eyes, why would it have been possible for you to have had blue eyes?
      (6) How does the Mendelian Law explain your taking after your parents more than after your grandparents, your grandparents more than after your great-grandparents, etc.
      (7) How much chance is there for any given mother to have all her four children be girls?
      (8) Why are identical twins identical, as compared to fraternal twins who are not?

   b. What does Darwin's theory of natural selection have to do with the way you are?
      (1) Report on Darwin's theory of natural selection.
      (2) Explain how the development of the human embryo illustrates evolution.
      (3) Give four other arguments for evolution, including one based on the structure or on parts of the human body.

   c. How is it possible for you or for anybody to have a characteristic that none of your parents, grandparents, or ancestors had?

   d. Can you name anything in language, science, mathematics, art, society, or life in general that does not have an opposite? Show how we can understand the nature of the following in terms of their opposites:. the structure of the atom, the structure of molecules, magnetic properties, lightning, a jet-propelled or regularly propelled plane, the nature of light, the process of walking, the argument of heredity versus environment, sex, and industrial relations.

   e. Report on the work of one of the following: Mendel, Lamarck, Darwin, DeVries, Galton, Weissman, or Burbank.

3. What personality differences or disturbances arise from differences or disturbances on lower levels?

   a. With two such examples as the thyroid and pituitary gland, describe how variations or changes in glandular activity change human behavior.

   b. How does the eunuch illustrate changes in appearance, mannerisms, and personality by chemical changes?

   c. What are the physical, chemical, and psychological aspects of fatigue?

   d. Explain how the mind can be considered a tool of the body and the body a tool of mind.

   e. Ask your doctor to explain the meaning of psychosomatic medicine and to give you an illustration of it. What are faith cures?

   f. Visit the occupational therapy department of a hospital or other institution, and explain the principle behind such a department.

   g. Explain the following: "During emotion there are physiological changes which affect the function of the intellect."

   h. Explain how a brain tumor could result in criminality.

   i. Take five diseases as polio, cancer, high blood pressure, tuberculosis, and diabetes, and show how they can affect activity on the various as chemical, physical, physiological, and psychological levels of the body.

   j. After seeing a demonstration of the lie detector, explain the physical, physiological, and psychological aspects of this device.

   k. Give three ways each in which your appearance, attitudes, interests, habits, mannerisms, and appreciations would be different if you were of the opposite sex.

4. How does your environment shape you?

   a. Answer the following:

     (1) In what ways would your thinking, values, attitudes, and reactions be different if you were born in France or in China rather than in the United States?

     (2) To what extent has the advance of science and industry since the Civil War changed the work, values, interests, lives, and culture of the present day?

(3) In what ways would your life, thinking, attitudes, values, and interests be different if you had been born into a Pygmy tribe?

(4) To what extent does your economic and financial status determine your social attitudes?

b. List as many ways as you can in which your home environment has patterned you.

c. List the ways in which your living in your community rather than elsewhere has patterned you.

d. List ways in which particular friends have influenced you.

e. In what ways has schooling conditioned you?

f. How would you have changed after you had lived for a year along the equator?

g. What traits did you develop at the birth of a younger brother or sister?

h. How can marriage change your life, thinking, attitudes, and habits?

i. How would prolonged discrimination on racial, religious, or political grounds affect you?

j. In what ways would you be different if you had lived all your life on a farm?

k. Describe five other factors of social environment that could change you.

l. Enumerate separately some of your physical, mental, and personality characteristics. Indicate which of these you think are the result of heredity, of environment, and those that reveal both.

m. Which way do we have a better chance for a healthier, happier world, through controlling heredity or controlling environment? What is the meaning and significance of cultural selection?

SUMMARY OF MICROCOSM INTERRELATEDNESS

Each of the above levels of activity involves a world of its own at increasing levels of complexity and yet is contained in an ever larger background framework of activity to make all of them interrelated systems within systems or, if you prefer, worlds within worlds, in which we fit as more conscious microcosms.

The following represents levels of interrelated microcosms comprising human beings and their surroundings, from minute quarks (i.e., the building blocks within electrons, protons, and neutrons) to the cosmos:

Quarks (in variety)
↓↑
Within electrons, protons and neutrons
↓↑
Within atoms
↓↑
Within molecules
↓↑
Within cells
↓↑
Within tissues
↓↑
Within physiological systems such as the digestive, reproductive, and nervous systems
↓↑
Within a biological organism
↓↑
Within a personality
↓↑
Within a society and culture
↓↑
Within international organizations
↓↑
Within natural surroundings
↓↑
Within a cosmos.

As we proceed from simpler to more complex levels, each successive new level becomes an immediately related background frame to the previous entity.

Arrows point in both directions as events at any given level can affect events in simpler as well as more complex levels, since all levels are interrelated. As the great German writer Goethe (1749–1832) expressed it: "In nature we never see anything isolated, but everything in connection with something else which is before it, beside it, under it, and over it."

The above, from quark to cosmos, are designations of interrelated worlds as devised by the human mind. Three general classifications exist in the above: the minute, the intermediate, and the large, all relative to the human organizer in the intermediate group. Intelligent microbes would have different classifications relative to themselves. And yet even with the above three classifications and their physical, biological, psychological, and social aspects, we assume one cosmos exists. We call that cosmos to order on our own terms.

Each of the above microcosms is an entity, a world of its own, even if not an isolate. Isolates do not exist because entities interact and form new entities. Electrons electrically interact with protons to form atoms. Atoms electro-chemically interact to form molecules. Molecules form cells, protoplasm (important in understanding living things), and tissue; tissues, organs; organs, bodies. From bodies emerge psyches that form societies and cultures which interact in and with surroundings that are parts of a huge cosmos. And because of the interaction of parts and the emergence of new world systems, the whole does not equal the sum of its parts. Hydrogen and oxygen atoms can combine chemically to form water. A water molecule has much different properties than its hydrogen and oxygen components.

DYNAMIC ASPECTS OF HUMAN PERSONALITY

The following topics listing dynamic aspects of human personality specify further suggestions for investigating the human microcosm:

1. physiological units and systems:
   the cell and protoplasm
   tissues
   organs
   physiological systems
2. the senses: adaptation, knowledge, and illusions
3. photosynthesis, nutrition, and self-preservation
4. metabolism
5. endocrine glands
6. emotions, glands, and nervous systems
7. electro-decephalography (separation of parts of the brain by electrical means))
8. genes, heredity, and environment
9. drives, habits, attitudes, ideals, intelligence, and other personality factors
10. Freud and psychoanalysis
11. anthropology, Gestalt psychology, and social conditioning
12. psychosomatic medicines.

Let me conclude a detailed interdisciplinary chapter such as this with a brief general suggestion as to team writing or team teaching on so integrative a topic as "Images, Ideas, and Evidences in a Study of Man."[3]

Basic Premises

1. Man is a conscious part of nature interacting with the rest of nature whether in everyday life, science, the arts, or government. Through examining our interactions and our products we may better know ourselves.

2. The human personality exists at various levels of activity that may be designated as subatomic, chemical, physical, biological, psychological, social, and cultural.

3. Activities at these various levels are interrelated, and variations in activity at one level affect activities at other levels. These activities at various interrelated levels afford mechanisms adaptive for change in a changing world.

4. Educationally, the more specialization is necessary to avoid superficiality, the more synthesis is needed to avoid provincialism. Work long enough at the elephant's ear and the elephant's ear becomes the elephant.

Main Concerns

1. Assumptions of a study or science of man (i.e., a study of ourselves)
2. Conceptual and experimental foundations for a study of man
   a. Foundations within disciplines
   b. Interdisciplinary concepts and experiences
3. Synthesizing insights and evidence
4. An interaction theory of knowledge and of behavior
5. Ecological interactions at various levels
6. Techniques and processes in a study of man
7. Synthesizing the great synthesizers of various cultural areas as shown below:

Synthesizing Synthesizers

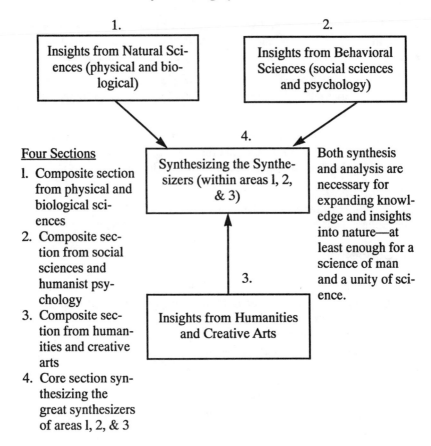

Figure 21.1
Ideas and Evidences in a Study of Man

NOTES

1. Kevin W. Kelley, ed., *The Home Planet* (New York: Addison-Wesley, 1988).

2. James S. Perlman, "Integration in College Courses in Science for General Education," *Science Education* 35, no. 2 (March 1951). The material in this chapter has been extracted from a unit on "Your Human Microcosm," originally written in a University of Minnesota interdisciplinary science program in 1948.

3. Taken from an interdisciplinary proposal I drafted for faculty at San Francisco State University in 1973.

SUGGESTIONS FOR FURTHER READING

Bohm, David. *Wholeness and the Implicit Order.* London: Routledge and Kegan Paul, 1981.

Boulding, Kenneth. *The Organizational Revolution.* New York: Harper & Row, 1953.

Davidson, Mark. *Uncommon Sense, The Life and Thoughts of Ludwig Von Bertalanffy.* Los Angeles: J. P. Tarcher, 1982.

Fromm, Erich. *The Revolution of Hope: Toward a Humanized Technology.* New York: Harper & Row, 1968.

Gore, Al. *Earth in the Balance.* New York: Houghton and Mifflin, 1992.

Graubard, Mark, *Two Fossil Sciences, Astrology and Alchemy.* New York: Philosophical Library, 1953.

22

Systems within Systems: Structural System Similarities

It is necessary to study not only parts and processes in isolation, but also to solve the decisive problems found in the organization and order unifying them, resulting from dynamic interactions of parts, and to recognize the behavior of parts different when studied in isolation or within the whole.
Ludwig von Bertalanffy, *Perspectives on General System Theory,* 1955

Let Professor von Bertalanffy provide a fuller context for the words above:

Problems of organization, of wholeness, of dynamic interaction are urgent in modern physics, chemistry, physical chemistry, and technology. In biology, problems of an organismic sort are everywhere encountered. It is necessary to study not only isolated parts and processes, but the essential problems are the organizing relations that result from dynamic interaction and make the behavior of parts different when studied in isolation or within the whole. These parallel developments in the various fields are even more dramatic if we consider the fact that they are mutually independent and largely unaware of each other.[1]

My illustration of an interdisciplinary program based on the human microcosm, described in chapter 21, is another case in point.

There are theoretical as well as experimental physicists. Therefore, why not theoretical biologists? Certainly Ludwig von Bertalanffy (1901–1972) was such and more: He was a pioneering promoter of the organismic approach to biology and a founder of general system theory. Again let me quote from Professor von Bertalanffy's *General System Theory*: "Systems theory is a broad view which far transcends technological problems and demands a reorientation that has become necessary in science in general and in the gamut of disciplines from physics and biology to the behavioral and social sciences and to philosophy. It is operative with varying degrees of success and exactitude in various realms, and heralds a new worldview of considerable impact."

PIECES AND PATTERNS

The whole does not always equal the sum of its parts, especially because of interaction among the individual parts. For example, four ordinary hydrogen atoms total 4.0330^2 atomic mass units (a.m.u.). When these four fuse to become a helium atom, the total a.m.u. *reduces* to 4.0028, the a.m.u. of a helium atom: a loss of .0302 a.m.u. This loss is due to interaction of parts in the fusing process in which mass loss takes the form of escaping energy as in the sun's radiation ($E = mc^2$). (See chapter 16). Suppose an oxygen atom O is properly added to the water molecule H_2O to get H_2O_2. What results is not another oxygen link to a chain of two hydrogen atoms and an oxygen atom of water, but a new hydrogen peroxide molecule, due to the interaction of parts. This illustrates the necessity for thinking in patterns rather than in sums of isolated pieces and for recognizing causes in a net of correlations.

COMMON CHARACTERISTICS OF SYSTEMS

Von Bertalanffy emphasizes seven basic principles of systems as systems, that is, as organizational similarities characterizing systems, regardless of their nature, size, or complexity. The common similarities are (1) openness of systems, (2) dynamic steady state, (3) negative entropy (system building organization), (4) interaction of parts, (5) equifinality and teleology, (6) irreversibility of processes, (7) cybernetic feedback, and (8) chance versus causality considerations. Let us briefly discuss each of these similarities.

OPEN SYSTEMS VERSUS CLOSED SYSTEMS

One of the main objectives of this book has been to depict scientists, both ancient and modern, as *open systems* operating at interrelated internal levels in interaction with their surroundings. The interaction characteristic of scientific investigation is diagrammed in chapter 11, Fig. 11.A and B. But more important, the diagram illustrates *openness of system* as articulated in considerable detail in chapter 11. Besides, all living things, including scientists *as humans,* exhibit dynamic open systems, whether in absorbing and emitting survival materials such as oxygen, carbon dioxide, water, and food, or else by information, impressions, and activities sent from and to their surroundings.

Open systems contrast with the closed systems characteristic of classical physics and chemistry and their assumptions, concepts, models, and experimental practices. Conventional sciences and their specializations have been very successful until recently by reducing complexities to elemental parts and understanding the whole primarily through its reduction into parts. As von Bertalanffy emphasizes, however, overlooked is the "dynamic interaction of parts."

Traditional physics and physical chemistry have been based upon closed systems called *isolates,* so called because they are assumed to be isolated from their surroundings. A number of chemical reactants are brought together in a closed vessel properly insulated or isolated from surroundings, and controlled apparatus gives information about the inside reactions, the rates of these reactions and chemical equilibria. In general, thermodynamic laws of physics and physical chemistry assume closed systems for their equations. And in closed systems, a particular quantity called *entropy,* used in the famous second law of thermodynamics,* increases to a maximum, to bring the process to its equilibrium in terms of random motion of the molecules involved. That means maximum disorder. But all living organisms, including component cells, are basically *open* systems that involve input from and output to environment. The second law of thermodynamics falls short, therefore, under actual, natural conditions. Living organisms, by interaction with their surroundings, by maintaining themselves, by building up and breaking down components, are not in a chemical and thermodynamic state of equilibrium, but in a more dynamic "steady state" associated with metabolic, chemical processes within living cells.

Von Bertalanffy agreed that insights had been gained in experimentally blocking off sections of nature in reductionist, mechanical approaches in physical, biological, psychological, and social realms. As a biologist, however, von Bertalanffy emphasized that specialized science could not explain complex interactions within a living cell; this underscored the necessity for a general systems theory that seeks and applies universal structured principles applying to systems in general. That means a new integrative discipline based upon apparent structural similarities or isomorphisms† in various fields.

---

*The second law of thermodynamics emphasizes that heat flows from a higher to a lower temperature, but not from lower to higher. Since all physical and chemical processes involve friction, the energy of any system over a period of time is reduced by lost heat, energy, or random motion of molecules called "heat death." The accumulation of such universal heat loss from organized to a randomized form is called "entropy," which is a measure of disorder at molecular and atomic levels.

†Objects or systems are isomorphic when they have identical or similar structure, form, or shape, and, of course, all kinds of human culture and creativity are pockets of anti-entropy, organization, and system building.

Organizational Aims of the General System Theory

Von Bertalanffy summarized the aims of general system theory as follows:

a. There is a general tendency toward integration in the various sciences, natural and social.
b. Such integration seems to be centered in a general theory of systems.
c. Such theory may be an important means for aiming at exact theory in the nonphysical fields of science.
d. Developing unifying principles running "vertically" through the universes of the individual sciences, this theory brings us nearer to the goal of the unity of science.
e. This can lead to a much-needed integration in scientific education.

Erwin Laszlo,[3] a close disciple of von Bertalanffy, lists four organizational principles of systems theory:

1. Natural systems as wholes have irreducible properties.
2. Natural systems maintain themselves in a changing environment.
3. Natural systems create themselves in response to the challenge of the environment.
4. Natural systems are coordinating interfaces in nature's hierarchy.

Entropy versus System-Building

Entropy has been discussed in chapter 20 and in a previous section of this chapter. Two universal tendencies have been noted: (1) the running down of the universe in its "heat death" and (2) the nuclear buildup in the sun and, most likely, other stars from simpler to more complex atoms. "Negative entropy" is a term now used to cover system *building*.

At present, for example, with Norbert Wiener's cybernetics and computer developments[4] and with general scientific system building, we find *negative entropy*, that is, organizational buildup in antithesis to entropic breaking down of systems. Unless, of course, we destroy ourselves by the dangerous gap existing between our technological advances and our lagging social intelligence. Are we helping "heat death" or chaos to win here on earth? Again, as Wiener expressed it, "We have modified our environment so radically that we must now modify ourselves in order to exist in this new environment. We can no longer live in the old one."[5] Do we humans have what it takes for social reorganization in line with technological advances?

Under conditions of lower temperatures and pressures, atoms of various elements combine to form molecules or compounds of increased complexity and organization, even to the extent of single-celled plants or animals evolving. Simple forms of life have developed into more complex forms in accordance with Charles Darwin's suggestions of individual variation and natural selection, and of Hugo De Vries's original insight of mutations. In photosynthesis, green plants use the sun's rays to form starch from water and carbon dioxide. And now here are we humans composed of complex systems within systems that integrate into the marvel of human awareness that we *are* aware. And certainly life at any level has tendencies to survive, to multiply, and to organize.

INTERACTION

We have seen, in previous chapters, an increasing emphasis on an interaction concept. Einstein's special relativity theory successfully related measurement of space, time, and mass to the relative velocity between the observer and the observed. Different relative velocities between the two reference frames change measurement values. Heisenberg's uncertainty principle made clear that light needed for observation imparts energy to what is observed so that the observer cannot simultaneously obtain an exact reading for the position of so minute a subatomic particle as an electron and its velocity when impacted by light. Again, here is a problem of interaction between observer and observed: the observer is intervening. Then there is Wiener's cybernetic feedback concept to illustrate an informational and learning interaction between observer and observed that shows the scientist to be an interactor rather than be a truly detached observer.

EQUIFINALITY AND TELEOLOGY

Equifinality is a principle characteristic of open, steady-state systems whereby the same final state of a process may be reached from different initial conditions and in different ways. Two mountain climbers may start at opposite sides of a peak and meet at the top. Of course, that introduces a goal, a purpose, teleology in a natural process. Or identical twins result from the division of one ovum. Thanks to the genetic regulation, even where such twins have been separated for some time and raised in different environments, the same basic traits can be recognized mentally, emotionally, and otherwise. We may compare this to, say, a closed-system chemical equilibrium where initial concentrations of reactants determine the final concentration of the products. Changing the initial closed-system conditions or the process seriously changes the final state.

Teleology does not contradict equifinality. All living organisms, including humans, to satisfy survival needs, show aspects of goal-seeking and purposefulness. While future goals determine present and future thought and behavior, they also provide different ways and conditions for reaching the same final state, as in the case of the two mountain climbers.

CYBERNETIC INFORMATION AND FEEDBACK

This was discussed and illustrated in chapter 20. We will merely add here that when open systems are self-regulating, they are cybernetic.

CAUSALITY VERSUS CHANCE IN NATURE

Chapter 19 adequately covered considerations of causality versus chance in nature for our purposes here. Both may be involved in system building. By mutation or other chance occurrence, a new species of life may arise. If surroundings are favorable, natural selection will take over and be causal in determining future developments of that new species.

TOWARD A UNITY OF SCIENCE AND A SCIENCE OF MAN

In line with his interdisciplinary and humanistic outlook, von Bertalanffy makes a strong appeal for a unity of science and a science of man. Let us summarize this in von Bertalanffy's own words:

> We come, then, to a conception which in contrast to reductionism, we may call perspectivism. We cannot reduce the biological, behavioral and social levels to the lowest level, that of the constructs and laws of physics. We can, however, find constructs and possibly laws within the individual levels. . . . *The unifying principle is that we find organization at all levels.* The mechanistic worldview, taking the play of physical particles as ultimate reality, found its expression in a civilization which glorifies physical technology that has led eventually to the catastrophes of our time. Possibly the model of the world as organization can help to reinforce the sense of reverence for the living which we have almost lost in the last sanguinary decades of human history. [That is, the unity of science is to be attained by recognizing structural similarities at different levels in sciences and in society.]
>
> Thus, system theory should prove an important means in the process of developing new branches of knowledge into exact science—i.e. into systems of mathematical laws. Corresponding conceptions and laws appear inde-

pendently in different fields of science causing remarkable modern paral-
lelism. Thus, concepts such as wholeness and sum, mechanization, central-
ization, hierarchical orders, stationary and steady states, equifinality, etc., are
found in different fields of natural science, as well as in psychology and sociology. . . . Reality, in the modern conception, appears as a tremendous *hierarchical* order of organized entities leading in a superposition of many levels from physical and chemical to biological and sociological systems. Unity
of Science is granted, not by a utopian reduction of all sciences to physics and
chemistry, but by the structural uniformities of the different levels of reality.[6]
(Italics added)

In chapter 21, I illustrated a hierarchical order from a simplest level of
quarks to human personality and worlds beyond, but I also reversed the direction from most complex levels to the simplest. This was to show that things
getting out of balance at simpler levels, say chemical or hormonal, can affect
what transpires in attitudes and behavior. The vertical arrows work both ways,
from simple to complex, and complex to simple.

Toward an Interaction Theory of Knowledge: Creativity

I heartily endorse von Bertalanffy's appeal for a unity of science and a science
of man; I also appeal for an interaction theory of knowledge. Scientists and all
human beings can be seen as microcosms of interactive open systems within
systems, internally and externally (chapters 21 and 22). Cybernetic feedback
is basic to scientific inquiry, as are considerations of nature and knowledge as
expansively open-ended.

We agree with von Bertalanffy that " any organism is a system, that is, a
dynamic order of parts and processes standing in mutual interaction."[7] This
applies to the scientific observer as a system as well. The observer is *dynamic*
in interplay, not passive. This is basic in sensory creativity, as seeing a rainbow; in projecting frames of reference when observing; and in interpreting
through devised and revised conceptual models, hypotheses, and theories and
attendant concepts, patterns, and associations. Besides, "our normal world is
shaped by emotional, social, cultural, linguistic and similar factors amalga-
mated with perception proper."[8]

That brings us to preliminary considerations of an interaction theory of
knowledge. In the previous chapters, *interactionism* showed itself in the development of science in different ways. It became clear that the properties of
things are not to be understood simply in terms of things themselves, but as
products of dynamic interaction between the observer and the observed. Your
weight on earth, under Newton's law of gravitation became an interaction

between your mass and that of the earth as measured from the two centers of mass. In Einstein's special relativity, measurement of any mass depended upon the relative speed between the reference frame of the mass measured and that of the observer. This applied also for space and time measurements. Why, then, should not all concepts of the sciences be examined further as properties of observed-observer interaction rather than merely as properties of things themselves? That requires another book with a leading edge in that direction.

NOTES

1. Ludwig von Bertalanffy, "General System Theory," New Rochelle, N.Y., The Center for Integration and Education; reprint in *Main Currents in Modern Thought,* 1955, p.71.
2. On the basis of oxygen set as 16.0000 a.m.u.
3. Erwin Laszlo, ed., *Perspectives of General System Theory* (New York: George Braziller, 1972). This is recommended reading for special applications of general system theory to various fields such as biology, physiology, economics, communication systems, humanistic psychology, ecology, and noetic planning by various interested writers in these fields.
4. Norbert Wiener, *The Human Use of Human Beings* (New York: Avon, 1970), chapter 9: "The First and Second Industrial Revolutions."
5. Ibid., pp. 249–50.
6. Ludwig von Bertalanffy, *General System Theory* (New York: Braziller, 1972), p. 49.
7. Ibid., pp. 86–87.
8. Ibid., p. 218.

SUGGESTIONS FOR FURTHER READING

Ashby, W. Ross. *Design for a Brain.* New York: John Wiley, 1952.
Bertalanffy, Ludwig von. *Perspectives on General System Theory.* New York: George Braziller, 1975.
———. *Robots, Men and Minds.* New York: Braziller, 1967.
Laszlo, Erwin. *Introduction to Systems Philosophy.* New York: Gordon & Breach, 1972.
Main Currents in Modern Thought. Retrospective Issue, November 17, 1940–November 17, 1975, nos. 2–5.
Mumford, Lewis. *The Myth of the Machine.* New York: Harcourt Brace, 1970.
Popper, Karl R., and John Eccles. *The Self and Its Brain.* New York: Springer-Verlag, 1977.
World Institute Council. *Fields within Fields . . . within Fields, The Methodology of Pattern.* New York: Julius Stulman, Pub., 1972.

Index

Acceleration: law of, 132–33; and laws of nature, 271–78; and principle of acceleration, 275–76

Adams, Herbert, 292–93

Adams, John C., 147–48

Agricultural Revolution, 31–32

Agriculture, in Egypt, 32

Alcymaeon of Croton, 45

Alexandria, library of, 83

Algebra, 39

Almagest (Ptolemy), 75, 83

Alphabet, 38

Anaxagoras, 53–54

Anaximander, 47, 165

Anaximenes, 47

Anderson, Carl D., 212

Andromeda galaxy, 152

Antigravity, 212

Aquinas, Thomas, 83

Aristarchus of Samos, outline, failure of sun-centered theory of universe of, 84

Aristarchus, 77

Aristotle, and causality, 165; on celestial motion, 73; on force, 131; on gravity, 115–16, 117; on light bending in water, 182; on matter, 51–53; on rainbow, 219; and theory of the universe, 71

Artemidorus, 222

Astrology, 41

Astronomy: in Babylonia, 41; and frames of reference, 65; and geometry, 71, 96; Plato on, 72; theory of, as cause of universe as machine, 154

Atomism, 45

Atoms, 206, 214

Babylonia: astronomy/astrology of, 41; calendar in, 40; mathematics in, 39; medicine in, 38; metal technology in, 36; science in, 34–43; urban revolution in, 33–34

Bacon, Francis, 57

Bacon, Roger: on the rainbow, 218 (table), 224–32, 242; on refraction, 225–26; scientific method of, vs. Newton, 233

Barrow, Isaac, 130

Bertalanffy, Ludwig von, 323; theories of, 343–50 (chapter 22); and unity of science, 348

Bessel, Friedrich Wilhelm, 89, 113

Bethe, Hans, 177

Bogen, Dr. Joseph, 200

Bohm, David, 307–8

Bohr, Niels, 206–7, 211–12, 298–99; complementary principle of, 302–3; correspondence principle of, 301–2, 307

Bolyai, John, 280

Bradley, James, 247–48

351